Global Bioconversions

Volume I

Editor

Donald L. Wise, Ph.D., P.E.
President
Cambridge Scientific, Inc.
Belmont, Massachusetts

CRC Press, Inc.
Boca Raton, Florida

Library of Congress Cataloging-in-Publication Data

Global bioconversions.

Includes bibliographies and indexes.
1. Biomass energy.   2. Biogas.   3. Biomass chemicals.
I. Wise, Donald L. (Donald Lee), 1937-
TP360.G57 1987   662'.8   86-20763
ISBN 0-8493-4507-3 (set)
ISBN 0-8493-4508-1 (v. 1)
ISBN 0-8493-4509-x (v. 2)
ISBN 0-8493-4510-3 (v. 3)
ISBN 0-8493-4511-1 (v. 4)

International Standard Book Number 0-8493-4507-3 (set)
International Standard Book Number 0-8493-4508-1 (v. 1)
International Standard Book Number 0-8493-4509-x (v. 2)
International Standard Book Number 0-8493-4510-3 (v. 3)
International Standard Book Number 0-8493-4511-1 (v. 4)
Library of Congress Card Number 84-11425
Printed in the United States

PREFACE

Global Bioconversions is a valuable reference text consisting of contributed chapters in which are described the most active bioconversion-based research and development projects around the world. The authors of these contributed chapters are those very pragmatic and dedicated technologists that are so well known for "getting the job done". This important text has as the major theme the bioconversion of organic residues to useful products. This major theme primarily encompasses the field of anaerobic methane fermentation. The text is intended to present the most practical and useful aspects of this emerging field of international significance. Due to the fact that bioconversion technology, with emphasis on methane fermentation, has been under development at key sites around the world, great care has been taken to include chapters from an international perspective. Further, as a perusal of the chapter titles will indicate, an emphasis has been made to address the important practical aspects of this global bioconversion work. It is to be noted that each chapter included in this text is the work of a praticular individual or group. There are no multiple chapters by more than one author or group. Thus, each of the included chapters most often reflects the dedicated career efforts of these workers. Further, each contributed chapter is presented on a stand-alone basis so that the reader will find it helpful to consider only the theme of each chapter. On the other hand, there is the unifying theme with all chapters of practical bioconversion systems on a global basis. A reader of this text, just entering the field, will find this text provides an excellent state-of-the-art presentation of the global import of bioconversion, with emphasis on methane fermentation. A reader of this text, who has experience in this field, will find the text to be essential for assessment and referral of this increasingly valuable technology.

THE EDITOR

Donald L. Wise, Ph.D., P.E., is Founder and President of Cambridge Scientific, Inc., Belmont, Massachusetts. Dr. Wise received his B.S. (magna cum laude), M.S., and Ph.D. degrees in chemical engineering at the University of Pittsburgh. Dr. Wise is a specialist in process and biochemical engineering as well as advanced biomaterials development. During his career he has managed a series of programs to develop processes for production of fuel gas, liquid fuels, and organic chemicals from municipal solid waste, an array of agricultural residues, and a wide variety of crop-grown biomass, especially aquatic biomass. Dr. Wise has also been primarily responsible for the initiation of development work on fossil fuels such as peat and lignite to gaseous fuel, liquid fuels, and organic chemicals, and he also originated work on the bioconversion of coal gasifier product gases to these products. Dr. Wise initiated a program to establish the engineering feasibility of converting large-scale combined agricultural residues to fuel gas by the action of microorganisms, a project ultimately involving joint effort with research workers in fifteen countries around the world.

Dr. Wise has worked in the area of biotechnology research and development for 2 decades, has approximately 50 publications in the field, and has edited a number of reference texts. As Associate Editor of *Solar Energy*, the journal of the International Solar Energy Society, he is responsible for the review of manuscripts in the biomass/bioconversion area. Dr. Wise is also on the Editorial Board of *Resources and Conservation*, an international journal published by Elsevier, Amsterdam. He has served as an international consultant in bioconversion for the United Nations and for the U.S. Agency for International Development (AID).

A meaningful portion of these programs that Dr. Wise initiated, and has been carrying out, is his meeting with experts across the U.S. and around the world, to become familiar with both current and practical aspects of bioconversion systems.

CONTRIBUTORS

Daniel Alkalay, Ph.D.
Chemical Engineer
Department of Chemical Engineering
Universidad of Tecnica Federico Santa
  Maria
Valparaiso, Chile

J. P. Blanchard, M.A.Sc.
Department of Food Science and
  Technology
Technical University of Nova Scotia
Halifax, Canada

Manuel J. T. Carrondo, Ph.D.
Assistant Professor of Biochemical
  Engineering
Department of Chemistry
New University of Lisboa
Monte Da Caparica, Portugal

Carlos Dosoretz M.Sc., Eng.,
Ph.D. Student
Department of Biotechnology
Migal Laboratories
Galilee Technological Center
Kiryat Shemona, Israel

M. Nabil Alaa El-Din, Ph.D.
Head of Research and Researches
Director of Biogas Project
Soils and Water Research Institute
Agricultural Research Center
Giza, Egypt

M. El-Housseini, Ph.D.
Researcher
Department of Agricultural Microbiology
Soils and Water Research Institute
Agricultural Research Center
Giza, Egypt

S. A. El-Shimi, Ph.D.
Researcher
Department of Agricultural Microbiology
Soils and Water Research Institute
Agricultural Research Center
Giza, Egypt

T. A. Gill, Ph.D.
Department of Food Science and
  Technology
Technical University of Nova Scotia
Halifa, Canada

P. J. F. M. Hack, Ir.
Engineer
Department of Research & Development
Paques, B.V.
Balk, Netherlands

Y. Z. Ishac, Ph.D.
Department of Microbiology
Faculty of Agriculture
Ain Shams University
Cairo, Egypt

Uri Marchaim, Ph.D.
Department Head
Department of Biotechnology
Migal-Galilee Technological Centre
Kiryat Shmona, Israel

Juan Oscar Mesa, M.Sc.
Mechanical Engineer
Department of Mechanical Engineering
Universidad Tecnica Federico Santa
  Maria
Valparaiso, Chile

Gerd Reinke, Dipl. Ing.
Diplom
Department of Mechanical Engineering
Universidad Tecnica Federico Santa
  Maria
Valparaiso, Chile

M. Ascensão M. Reis
Lecturer
New University of Lisboa
Monte Da Caparica, Portugal

Alejandro Saez, Ph.D.
Mechanical Engineer
Department of Mechanical Engineering
Universidad Tecnica Federico Santa
  Maria
Valparaiso, Chile

E. A. Saleh, Ph.D.
Lecturer
Department of Microbiology
Faculty Agriculture
Ain Shams University
Cairo, Egypt

Global Bioconversions

VOLUME I

The Use of Biogas in Diesel Engines, the Higher Performance
and Low Contamination Advantages
Methane Production by Continuous Digestion of Farm Wastes
Fixed Film Anaerobic Digestion
Biogas from Sanitary Landfill Technique in Egypt
The U.A.S.B.-Reactor Treating Paper and Board Mill Effluent
Bioenergy Recovery and Conservation in Israeli Agriculture — Anaerobic Digestion
of Cotton Stalks

VOLUME II

Thermophilic Anaerobic Digestion of Agricultural and Urban Wastes from Liquid to
Semi-Solid State
Fixed Bed Anaerobic Reactors for the Biological Treatment of Liquid Wastes
The Production of Chemicals and Fuels from Municipal Solid Waste
Effluents of the Food Industry in Mexico: Environmental Impacts on Soil and Water
Resources and Possible Solution Using the Biotechnological Approach — Case Problem:
The Corn Industry
Anaerobic Degradation of Municipal Solid Waste — Laboratory Scale Tests
RBC for Carbonaceous Removal and Nitrification of Municipal Wastewater
Utilization of Straw and Molasses as Feed

VOLUME III

Packed Bed Reactors for Wastewater Treatment
High-Rate Anaerobic Treatment of Industrial Wastewaters
Nitrogen Removal in Swine Wastes by Aerobic-Anoxic Steps
Organic Waste Stabilization Throughout Composting and Its Compatibility with
Agricultural Uses
The Anaerobic Digestion of Pig Waste in Four Different Reactor Designs
Depollution of Pig Waste Combined with Energy Recovery

VOLUME IV

Mixing in Biological Treatment Systems
Full Scale Anaerobic Digestion of an Animal Fat Containing Substrate in Sludge Blanket
Reactor and Fixed Film Reactor
Biogas Recovery Techniques from a Municipal Solid Wastes Landfill Undergoing
a Thermophilic Anaerobic Fermentation
Present State of Anaerobic Digestion for the Treatment of Animal Wastes in the
Catalonian Area (Spain)
Biogas Production from Rice Straw and *Clitoria ternatea*
Anaerobic Treatment of Potato Processing Wastewaters
Application of Solid State Fermentation to Utilize Solid Wastes

TABLE OF CONTENTS

VOLUME I

Chapter 1
The Use of Biogas in Diesel Engines, the Higher Performance and Low Contamination
Advantages 1
Alejandro Saez, Gerd Reinke, Daniel Alkalay, and Juan Oscar Mesa

Chapter 2
Methane Production by Continuous Digestion of Farm Wastes 41
J. P. Blanchard and T. A. Gill

Chapter 3
Fixed Film Anaerobic Digestion 101
Manuel J. T. Carrondo and M. Ascensao M. Reis

Chapter 4
Biogas from Sanitary Landfill Technique in Egypt 131
M. Nabil Alaa El-Din, E. A. Saleh, Y. Z. Ishac, S. A. El-Shimi, and M. El-Housseini

Chapter 5
The U.A.S.B.-Reactor Treating Paper - and Board Mill Effluent 145
P. J. F. M. Hack

Chapter 6
Bioenergy Recovery and Conservation in Israeli Agriculture — Anaerobic Digestion of
Cotton Stalks 155
Uri Marchaim and Carlos Dozoretz

Index 175

Chapter 1

THE USE OF BIOGAS IN DIESEL ENGINES, THE HIGHER PERFORMANCE AND LOW CONTAMINATION ADVANTAGES

Alejandro Saez, Gerd Reinke, Daniel Alkalay, and Juan Oscar Mesa

TABLE OF CONTENTS

I. Abstract 2

II. Introduction 3

III. Background 4
    A. Theoretical Approach 4
    B. Stationary and Vehicular Application 4
        1. General Considerations 4
            a. Purpose 5
            b. Energetic Replacement Ratio 7
            c. Engine Requirements 7
        2. Electric Power Generation 7
            a. Synchronized Operation 7
            b. Independent Operation 8
        3. Automotive Applications 8
    C. Biogas Treatment Methods 9
    D. Fuel Properties and Operation Requirements 10
        1. Heat of Combustion (Hi) 10
        2. Heating Value of Mixture 11
        3. Density 11
        4. Air/Fuel Ratio 11
        5. Sulfur Content 13
        6. Ignition Ranges and Temperatures 13
        7. Cetane Number (CN) 13
        8. Octane Number (ON) 13
        9. Laminar Flame Velocity 14
        10. Fuel Handling Safety 14
            a. Stationary Use 14
            b. Vehicular Use 15

IV. Application to Standard Diesel Engines 15
    A. Engine Modifications for Dual Fuel Operation 15
        1. Stationary Engines 15
            a. Intake System 15
            b. Gas Supply System 16
            c. Injection System Modifications 17
            d. Governors 17
            e. Safety Devices 17
        2. Automotive Engines 17
            a. Intake System 17
            b. Gas Supply System 17
            c. Governors and Safety Systems 18

    B.   Performance 18
        1.   Full Load Behavior 18
        2.   Influence of the Biogas Composition, Full Load 19
        3.   Influence of the Injection Point, Full Load 20
        4.   Cylinder Pressure Development 20
        5.   Summary of Performances Described 22

V.   Practical Experiences 23
    A.   Illapel Project 23
        1.   Introduction 23
        2.   Installation 25
        3.   Results 26
            a.   Long Term Operation Test 26
            b.   Fuel Cost Analysis 26
    B.   Thermophilic Digester/Diesel Engine, Total-Energy Pilot Plant 27
        1.   Plant 28
            a.   Fermentation Subsystem 28
            b.   Motor/Generator Subsystem 30
            c.   Thermal Control Subsystem 30
            d.   Operational Control and Data Acquisition
                    Subsystem 30
        2.   Mass and Energy Balances 31
    C.   Municipal Garbage Collection Truck 31
        1.   Biogas Cylinders 32
        2.   Motor Modifications in Test-Bench 32
        3.   Performance in Test-Bench 33
        4.   System Installation in Truck 33
        5.   Performance in City 33
        6.   Reduction in Contamination 33
        7.   Costs 33
        8.   Conclusions 33

VI.   Nomenclature and Abbreviations 34

References 40

I. ABSTRACT

Biogas generated in landfills or from agricultural wastes can be used with advantages in stationary and transport diesel engines. Technical aspects of runing diesel engines dual fueled with biogas are presented through work performed at the Technical University Federico Santa María (UTFSM), of Valparaíso, Chile, with respect to:

- Performance vs. gas substitution.
- Effect of fuel injection point on motor behavior.
- Mechanical and thermal solicitations of standard diesel engines.
- Effect of the former conditions on the pollution levels of the exhaust gases.

In this respect, some applications performed by the research group are presented, which include:

- Practical results of running garbage collection trucks on purified landfill biogas.
- The running of a conventional diesel engine/electrical generator on biogas from a 4000 head pig farm, supplying the required energy for the fodder preparation, animal rearing, and other alternative uses at the farm.
- A pilot plant constructed under the total-energy principle running with a biodigester operated thermophilically, at a small pig farm.

II. INTRODUCTION

The technology of the treatment of organic wastes has lately had an almost explosive growth. This has been initially due to the increase in the environmental conscience (late 1960s and early 1970s), and in a second stage due to the energy crisis.

On the other hand the biotechnology itself has had a very important development, partly because of the very same reasons, but also because of the better knowledge of the biochemistry and genetics of microorganisms. This way, new processes have been developed and old processes have been improved. Special importance have been acquiring the anaerobic processes, such as: sewage water treatments, biological treatments of industrial wastes, industrialized landfills, anaerobic fermentation of agricultural wastes, and others. Their particular advantage is that an energetic (biogas) is produced and its use as a fuel may become in many cases the main reason for the implementation of the process. Depending on the particular application of the process, a wide range of technological complexity of plants exists. Some are ready for or are already in production, while others are still in the developing stage.

Biogas has use as an energetic for direct combustion in heating applications and also for the running of internal combustion engines in replacement of traditional fuels.[1] In the latter sense, the firing of otto engines with gas, whether it were biogas, natural, or producer gas has had a long history and has been already developed to commercial application, under the name of Gas Engines. However, when it comes to the running of diesel engines with methane or directly with biogas, there has been very little done due to the ignition problems which are involved. This then, is the main subject of this work.

As background, the technical feasibility of an energetically integrated system is analyzed first. Then, the different applications to stationary and vehicular engines are studied. The fuel properties and purification methods available are mentioned, to give place to the diesel engines operation requirements.

In a second chapter the application of this technology to standard diesel engines is studied with the motivation of searching for a field of application at a minimal cost, without the need of special designs. Here, the engine modifications for dual-fuel running are described, including the control systems needed for this particular mode of operation. This is completed with the results on performance and other characteristics of operation, which have been studied at the UTFSM up to this date.

The final chapter describes the practical experience which the research group has had during 4 years of existence. These include:

1. Test installation, running and evaluation of a diesel motor/electric generator, operated at a pig farm which had already a 350 m³ industrial biogas facility.

2.   Design and construction of a thermophilic biogas/diesel engine pilot plant operating as a total-energy system.
3.   Development, installation, and test of a prototype of a garbage collection truck, run on landfill biogas.

The environmental impact of this technology is in this case twofold. First, the possibility of using one of the products of the process as an energetic may make the treatment of an offensive waste economically attractive. Secondly, the running of a diesel engine on biogas dramatically diminishes the soot and other emissions of contaminants.

III. BACKGROUND

A. Theoretical Approach

Among the different products of anaerobic biodigestion of organic wastes the one to be analyzed in this contribution is the use of biogas as an energetic, specifically as a fuel for diesel engines. Other industrial applications of this energetic are

- Boiler fuel
- Dryers and dehydration processes
- Other heating uses

The analysis of the use of this fuel in internal combustion engines includes otto and diesel cycle engines. However the current tendency in industrial engines is the diesel type, mainly because of better energetic efficiency. This includes, mass transportation of goods and people in large urban areas. Another major advantage where the engines run on gaseous fuels is the important diminishing of atmospheric contamination through exhaust gases.

In agricultural applications, even when the exhaust gases contamination has a lesser impact, the energy savings that may come with the implementation of biogas plants shows an attractive effect. This, in view of the increasing technification of the agricultural activity (specially animal rearing) which necessarily is energy intensive.

When it comes to the use of biogas in diesel engines, two different applications arise: one in stationary engines, mainly for electric or mechanical power generation; and the other in vehicular uses, either trucks, buses, or tractors. In stationary engines the energy flow analysis shows through the Sankey diagram, that an adequate design of the system (as a total-energy system) allows an optimal use of the energy content of the fuel. The residual heats from the exhaust gases and cooling water can be used as a thermal source, either for the digestion process itself and/or any other application. As an example, in Figure 1, the comparative analysis of the same diesel engine run on a biogas diesel fuel mixture, is made for the coupled operation with either a mesophilic or thermophilic digester.[2] The design and operating conditions of the two systems are shown in Table 1. Finally, a comparative diagram is shown of the system analyzed, where the usable energy appears for different biomass sources, and comparing otto and diesel engines (Figure 2). However, it has to be considered that not all the energetic performance of the diesel engine is applicable to the biogas, because 15% of the energy comes from the diesel fuel needed for ignition.

B. Stationary and Vehicular Application

1. General Considerations

The use of biogas in diesel engines presents characteristics which are essentially different from those in otto engines. The fact is that a special ignition procedure is needed because of the high ignition temperature and low cetane number of biogas. Several methods have been devised to ignite the gas: spark plugs, glow plugs, and others. The most practical one,

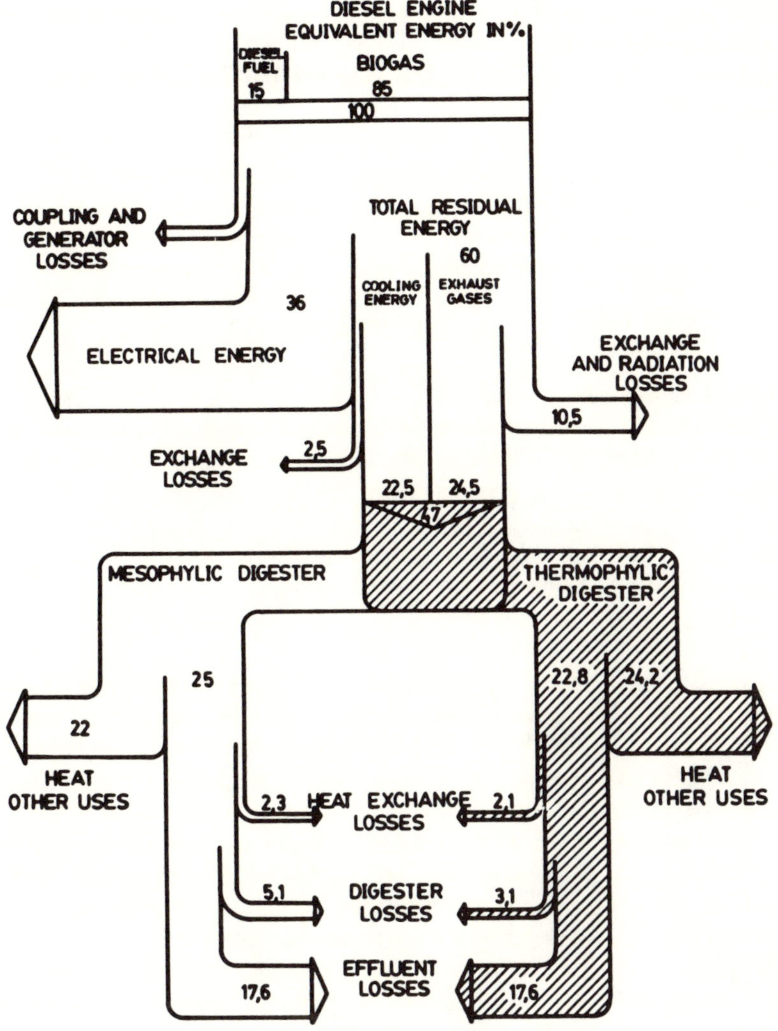

FIGURE 1.   Comparative Sankey diagram of diesel engine interconnected with a meso-
or thermophilic digester.

which is rapidly developing, is dual fueling with diesel. This principle is suitable to both
stationary and vehicular applications. The gas is mixed with the input air in the inlet manifold.
This mixture is compressed, but its final temperature is not high enough to produce combustion. Then, fuel oil is injected in the usual way and its combustion ignites the gas-air
mixture. The amount of fuel oil needed may be as little as necessary to ignite the mixture,
and as much as the particular application requires.

a. Purpose

For proper design and operation of any combined fuel system, its purpose must be clearly
stated. So far, there are basically three reasons for it.

1.   To reduce exhaust gas contamination.
2.   To save diesel fuel.
3.   To curb environmental noise.

Table 1
OPERATING CONDITIONS OF DIESEL TOTEM
WITH MESO OR THERMOPHILIC DIGESTER

	Mesophilic	Thermophilic
Operation temperature (°C)	35	60
Hydraulic retention time HRT (days)	15	7
CH_4 in biogas (%)	70	60
Volatile solids degrad (%)	30	35
Total solids (% m)	8	8
Volatile solids (% m)	6.4	6.4
Biogas produced (m^3/kg HBL · day)	0.0173	0.0202
Methane produced (kg/kg HBL · day)	0.0081	0.0081
Volume of digester (m^3/kg HBL · day)	0.018	0.0084
Heat loss of the digester (through walls and by water evaporation) (kJ/m^3 · day)	2090	3344

Note: In the thermophilic digester the effluent heat is recovered from 60 to 35°C, whether in the mesophilic it is all lost. The lower net losses in the thermophilic process are due to the smaller volume of the digester needed for the same amount of gas produced.

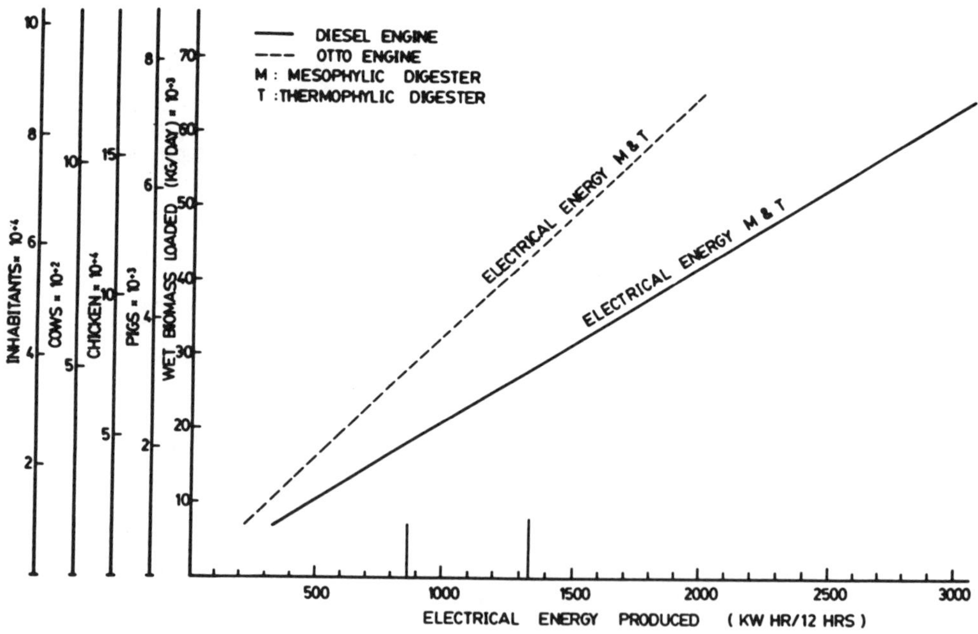

FIGURE 2.   Usable energy for otto or diesel engines run on biogas.

It is true that these advantages can be achieved simultaneously, but the first and last reasons are valid especially for vehicles in city traffic: local buses, delivery trucks, garbage collectors, taxis, etc.

One of the early works was conducted with municipal buses in Vienna.[3] In that case, LPG was employed, but the economical advantage is questionable, because liquefied petroleum gas (LPG) is not a renewable energy fuel and its cost is not always low. However, smoke reduction is significant, and that was certainly the main achievement of that program.

If the main purpose of the conversion is to economize diesel fuel, then it would be convenient to use a low cost gas in large amounts.[4] However, the actual cost of any gas depends strongly on local conditions, such as: production, storage, delivery, and taxation. LPG is widely available around the world, but biogas and compressed natural gas (CNG) compete in low cost under certain circumstances. For use in vehicles, biogas must be processed and stored in high pressure bottles or using cryogenic techniques, which adds to the original cost.

An attractive example of stationary engine application is electric power generation near the site of gas production: natural gas wells, biodigesters, sewage water treatment installations, garbage disposal plants, landfills, etc. In these cases, no compression or other processes are necessary. The gas can be fed directly to the engine at low pressure with a reasonably low cost.

b. Energetic Replacement Ratio

It is common practice to express the amount of gas as a percent of its energy content in relation to the total energy input. In other words, it may be understood as the fraction replaced by gas, of the total fuel required.

$$\psi = \frac{mfBG \cdot HBG}{mfBG\ HBG + mfD \cdot HD}\ 100(\%) \tag{1}$$

For environmental applications, only 30 to 40% gas is enough to achieve a considerable smoke reduction, but fuel savings may not be attractive enough.[5] For fuel saving, a higher percent of fuel replacement is desirable; over 80% substitution is possible to obtain when the engine is working at full power. However, some problems such as pressure peaks and injector obstruction are being investigated. For higher replacement and variable speed, an accurate regulation is more difficult and will be explained later.

c. Engine Requirements

When a standard diesel engine is adapted to dual fueling with biogas, some requirements should be observed. Each design has temperature limits which may be overrun causing serious damage, so careful experimental work can give the convenient operation conditions.[6] Also, not only the thermal stresses, but the mechanical ones should be evaluated too, mainly through the monitoring of the combustion chamber pressure. The usual compression ratios for diesel engines are in the permissible range for this application, producing acceptable pressure gradients and avoiding detonation occurence, but higher compression ratios must be treated with special care. Combustion chamber configurations can be of great influence in the performance and they should be checked for proper behavior. The type of injection system and governor must be taken into account for the design of the regulation devices, depending on the application.

2. Electric Power Generation

Fuel regulation requirements are different whether the generator is to be synchronized to the main electric supply or if it is going to work independently.

a. Synchronized Operation

This case is the simplest from the point of view of control and regulation.[7] The original injection pump and governor need not to be disturbed, except for the installation of a limiting device to reduce the diesel oil supply to the desired value. A manual control is sufficient for the gas, as the full power requirements are preset in advance. Also a conventional synchronizing panel is needed to connect the generator to the line.

b. Independent Operation

If the generator is going to work powering an independent load, the situation is different: it is simpler from the electrical point of view, but the gas regulation becomes more complex. The engine requires more important modifications, depending whether it is intended to work with variable or fixed fuel ratios. More details about the necessary modifications for either working mode are found in Section IV, A.

3. Automotive Applications

In stationary engines, the high content of carbon dioxide in biogas does not present any problem in the combustion chamber, except for a minor loss of output power. If the piping is sized, taking into account the higher volume required, the engine works satisfactorily, providing that it is not far from the gas source.

For use in vehicles, the gas must be processed, compressed, and stored in high pressure bottles; or cryogenic technics at low pressure have to be applied. All these add to the original cost of the fuel. The process consists mainly of removing the carbon dioxide in order to increase its energy content and therefore to reduce storing space and cost. In addition, it will be highly recommended to consider additional stages in the process to remove moisture and H_2S that may produce corrosion in the more complex equipment used in a vehicle. The fact that high pressure storage bottles are necessary with all their related accessories (pressure reducer, charging connection, etc.) means a more expensive and complicated equipment than in stationary engines.

Another additional complexity of vehicular application is the wide variation of speed and load applied to the engine of any vehicle with mechanical transmission. They often extend out of the normal range, from low speed overload to overspeed and braking action. The braking action can be considered a "negative load", because crankshaft torque is inverted and it occurs when the vehicle travels downgrade or decelerating with the throttle closed or partially open. Speed may fall in the normal range or if it exceeds the limit, it may lead to a dangerous overspeed condition.

The actual operation point is determined by a number of variables: vehicle weight, road and wind resistance, road grade, throttle position, and gear selected. The last two factors are under control of the driver and his particular driving habits. Vehicles with automatic transmission and/or torque converter are far less dependant of driver's habits and engine operation is therefore limited to a narrow zone. Figure 3 explains torque curves of a diesel engine equipped with a maximum/minimum governor, such as Bosch type RQ or equivalent.

We will consider engines with the following characteristics:

Cycle — 4 stroke
Intake system — naturally aspirated
Compression ratio — up to 19:1
Combustion chamber — direct injection
Injection pump — in line (Bosch type or similar)
Injectors — multihole, spring loaded (Bosch or similar)
Governor — mechanical (centrifugal) maximum and minimum type (Bosch RQ or
  equivalent)

Therefore, for safety and gas economy, the system must meet the following conditions. First, the gas input to the engine must be prevented if no diesel is being injected. Otherwise unburned gas will pass to the exhaust pipe with danger of explosion in the silencer, apart of the gas waste. Second, the gas flow must be prevented when the engine is not moving, moving slowly, or during starting. Otherwise, gas may come through the air filter out to the atmosphere, with the danger of explosion in the engine compartment. In other words,

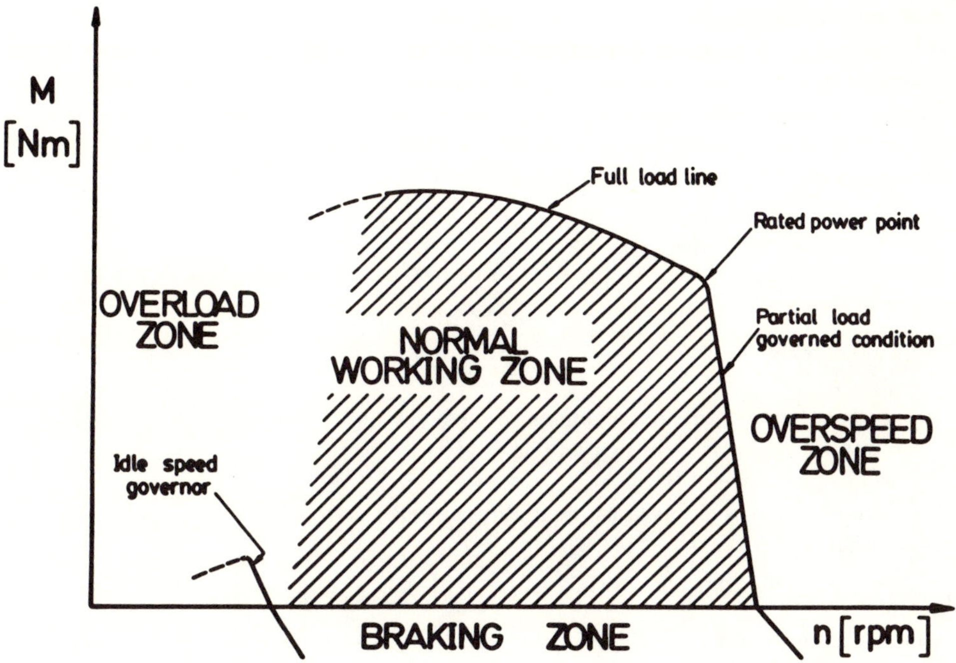

FIGURE 3.　Torque curves of diesel engine. (Dotted lines indicate overload).

gas should be delivered only when the engine is operating within the normal working zone as shown in Figure 3.

C. Biogas Treatment Methods

For the use of biogas in stationary diesel engines the gas treatment is not necessary, because of the following reasons.

- The stoichiometric air to biogas volumetric ratio usually being around 6:1, implies that the increase of inerts in the mixture due to the CO_2 is less than 1/12, when compared to pure methane firing.
- The fact that around 15% diesel fuel is injected for ignition and that values of $\lambda > 1$ are used, further minimizes the effect of the presence of CO_2 in the biogas.
- The low H_2S content in biogas, usually less than 0.2% (as H_2S, v), is significantly smaller than the tolerated sulfur content in diesel fuels, which according to Deutsche Industrie Normen (DIN) and American Society for Testing Materials (ASTM) standards ranges up to 0.7% (as S, m).

However, for vehicular use where compression of the fuel gas is necessary, the elimination of a 40% of inert CO_2 is convenient. The cost of compression, or cooling to cryogenic liquefaction levels, economically and practically justifies a previous gas treatment process. The small amounts of H_2S are simultaneously eliminated by most of the processes at no additional cost, with the extra gain in the deodorizing of the fuel gas.

Moisture traps have not to be forgotten. Most of the treatment methods available have been fully developed in the past in other applications and many different processes are available.[8-10] In a past CRC publication *Fuel Gas Production from Biomass,* Ashare presents a very complete and organized survey.[8] The different processes can be classified as

Table 2
GAS COMPOSITION OF "LA FERIA"
LANDFILL

Date	CH_4 (%)	CO_2 (%)	Air (%)	Others
May 1983	49	25	25	1
September 1983	58	35	7	–
June 1984	56	38	5	1

Chemical solvents
  Amines
  Monoethanolamines, diethanolamines, triethanolamines, diglycolamine, diisopropilam-
    ine, and mixtures with di- or triethylene glycol
  Potassium carbonate
  Different processes, licensed as: benfield, catacarb, vetrocoke, lurgy, and others
  Other chemical solvents
  Sulfinol, alkazid, etc.
Physical solvents
  One of the most economical and widely used is scrubbing with *water* at an intermedi-
    ate pressure (7 to 34 bar)
  Others are
  Fluor solvent, purisol, rectisol, selexol
Adsorption on solids
  Many commercial adsorbents suit the needs, among them: activated carbon, alúmina,
    and some molecular sieves.
Membrane separation
  Still not very popular, but with a promising future because of its compactness and easy
    operation
Distillation under cryogenic conditions
  When combined with liquefaction becomes very interesting, but it is still too expensive

After any of these processes the gas comes out with 95 to 100% methane, ready for
pipeline transportation, pressurizing up to 200 bar or even for cryogenic liquefaction ($-160°C$
and 2 bar) whichever procedure suits the particular use of the gas.

D. Fuel Properties and Operation Requirements

Biogas is a gaseous mixture mainly of methane (CH_4) between 50 and 70%v and CO_2
between 30 and 50%v. Other gaseous components may be present, but in small amounts.

The range of mixture depends on the biogas production system. Also the raw organic
material used to feed the biodigester and the temperature of operation affects the gas composition. As an example the composition obtained in the landfill "La Feria", Santiago de
Chile, is shown in Table 2. The properties which are relevant to the combustion in diesel
engines, compared to diesel fuel are shown in Table 3.

1. Heat of Combustion (Hi)

As the most important property in a fuel, this value changes with the methane content in
the biogas and with the substitution rate (Figure 4). The heating value grows with the
methane content. Special consideration has to be given to the fact, that at lower CH_4 contents
the increased CO_2 fraction occupies a larger portion of the gas volume, displacing part of

Table 3
FUEL PROPERTIES

	Diesel oil	CH$_4$	60% CH$_4$: 40% CO$_2$
Normal density (kg/m^3)	815—855	0.72	1.2
Stoichiometric air (kg/kg)	14.5	17.2	6.1
Boiling temperature (°C)	175—450	− 162	− 128
Heating value (inf) (MJ/kg)	43	50	17.5
Ignition range (% vol)	0.48—1.35	0.7—2.1	0.7—2.3
Cetane number	45—55		See Figure 8

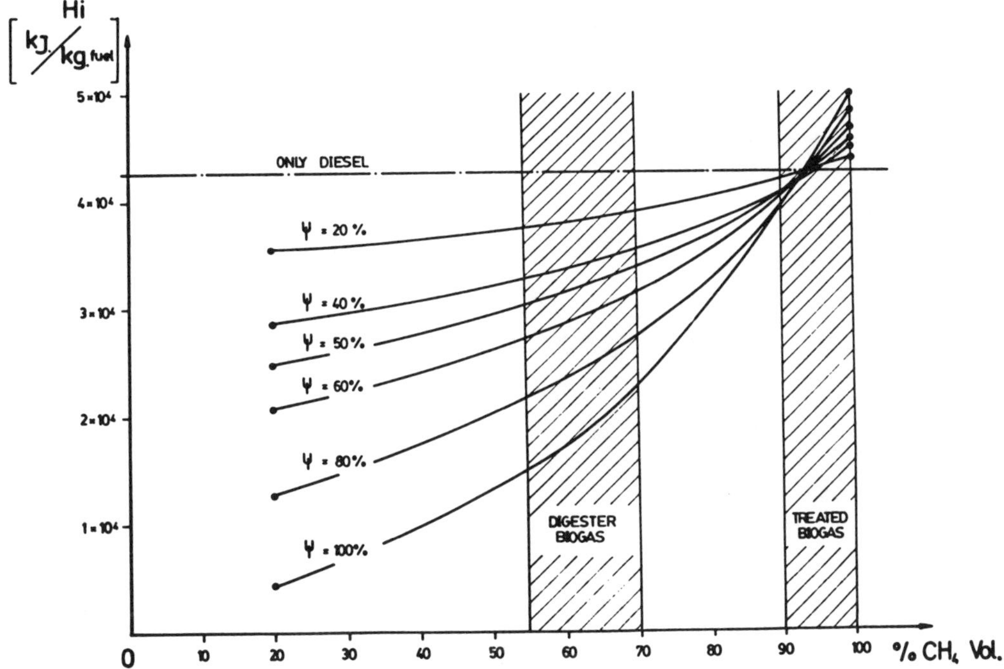

FIGURE 4. Dual-fuel heating value.

the air. This has a direct consequence on the A/F ratio (and λ), which in turn affects all the combustion conditions and the volumetric efficiency.

2. Heating Value of Mixture

For the mixture oxidant/fuel it is important to know the energy density, which varies as shown in Figure 5. Similarly to the previous point the CH$_4$/CO$_2$ ratio influences the heating value of the mixture in relation to λ, and becomes important for soot control.

3. Density

Biogas has a density which ranges closely around the air density, depending on the biogas composition (Figure 6). This variation has consequences on the volumetric efficiency and on safety regulations.

4. Air/Fuel Ratio

The stoichiometric air/fuel ratio is dependent on the substitution rate and the composition

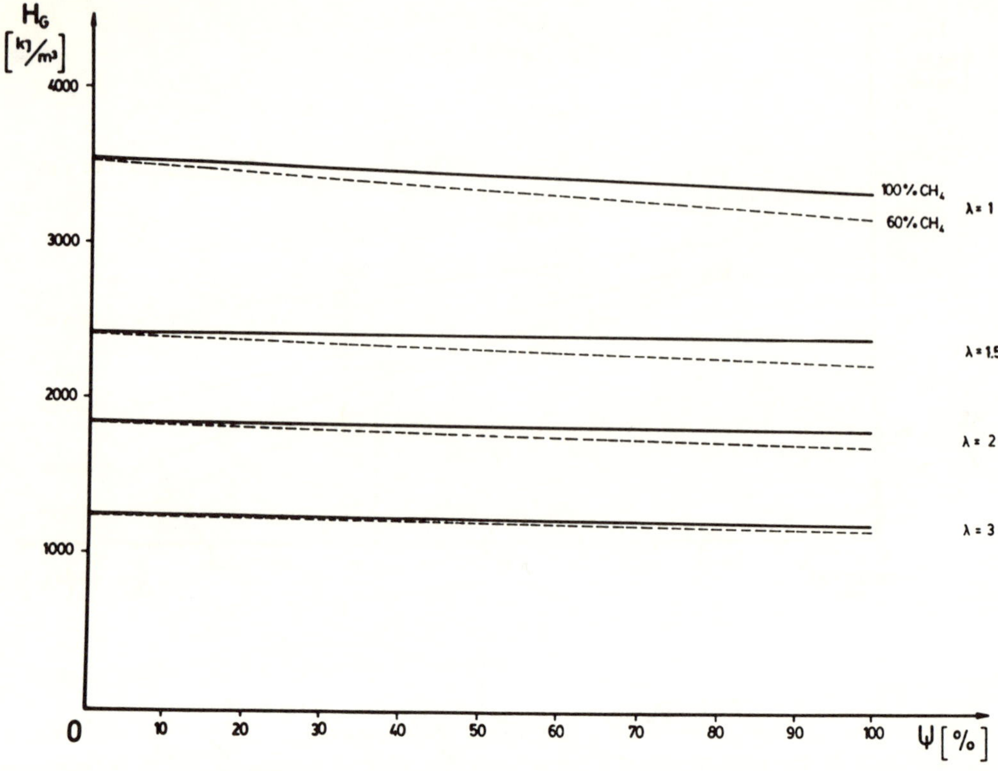

FIGURE 5.   Heating value of chamber gas mixtures.

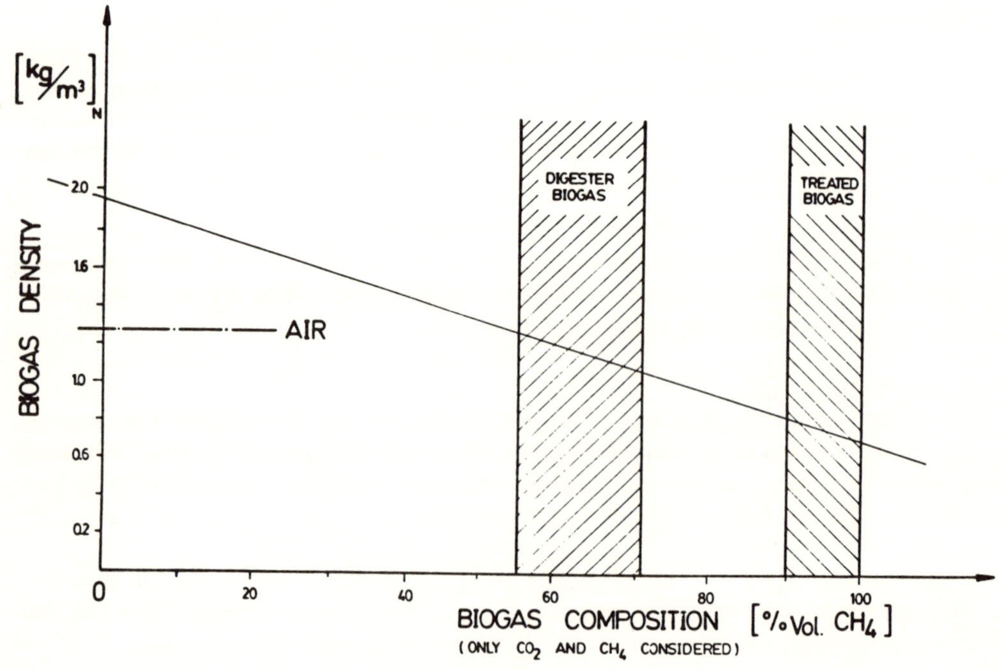

FIGURE 6.   Biogas density.

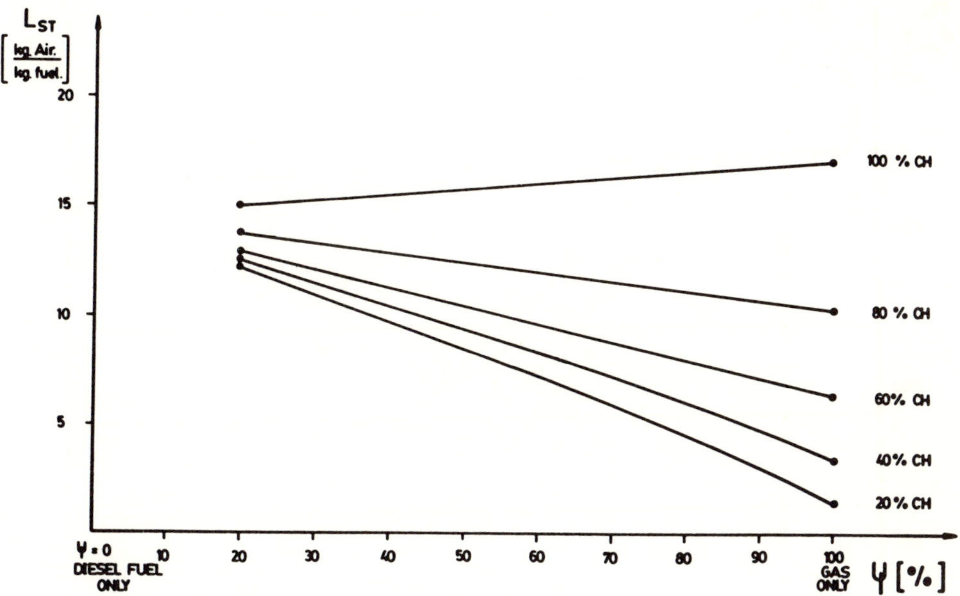

FIGURE 7.   Stoichiometric air/fuel ratio.

of the biogas. As *fuel,* it is usually considered the diesel fuel mass plus the total biogas mass (Figure 7). For pure methane it can be seen that more air is required; but the λ limitation for diesel fuel (λ = 1.3 to 1.5) probably may be overcome, getting closer to stoichiometric combustion, since the soot problem is less severe.

5. Sulfur Content

As mentioned before, the maximum sulfur content of untreated biogas is largely below the diesel fuel sulfur content. Consequently, from the fuel characteristics point of view no special considerations has to be given to it, since the lubricants for diesel fuel are formulated to tolerate higher sulfur contents. However, if TOTEM operation (or similar) is foreseen, the heat recovery exchangers must be adequately specified to avoid corrosion.

6. Ignition Ranges and Temperatures

These characteristics are greatly divergent. In air the high ignition temperature of methane, compared to the diesel fuel, implies difficulties to obtain the adequate firing condition by compression only. This is *the reason* for dual fueling.

7. Cetane Number (CN)

The cetane number is a function of the substitution rate and the methane content of the biogas (Figure 8). Low cetane numbers at over 50% of substitution will affect the combustion development and the pattern of the pressure-time curve. Specially, the ignition delay may become too high for a smooth diesel engine operation.

8. Octane Number (ON)

This characteristic is usually of main importance in otto engines and generally not considered for diesel engine operation. However in this case it may have a consequence in the out-of-time self-ignition of the gas mixture. Fortunately the high ON of methane (Table 3) is a safeguard towards the nonoccurence of this undesired phenomenon.

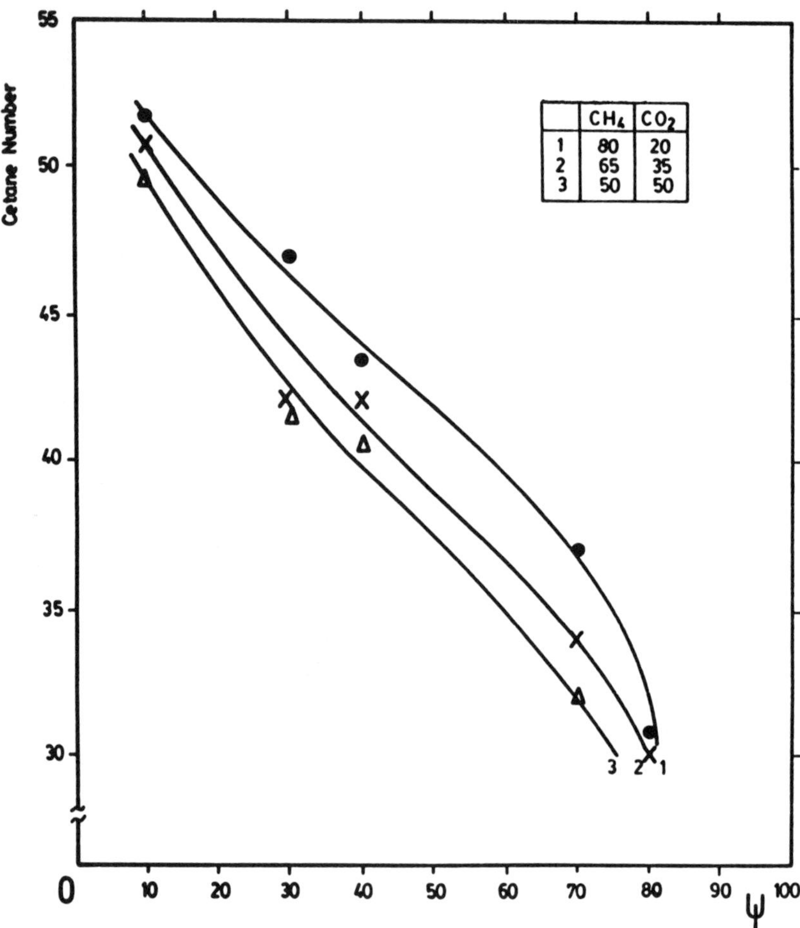

FIGURE 8.   Cetane number of diesel fuel biogas mixtures.

9. Laminar Flame Velocity

The propagation of the flame front affects substantially the burning of the mixture. Being that it is very difficult to know the actual turbulent flame velocity in the engine chamber, the laminar flame velocity will be used as reference (Table 4).

10. Fuel Handling Safety

Two types of biogas as a fuel are to be distinguished, from the point of view of its use and properties.

a. Stationary Use

In this case it is used as it comes from the digester or other source, with no, or minor treatments. For handling, it can be considered similar to any gaseous fuel-inert mixture, having consideration of the fact that it may be saturated in humidity and that some H_2S may be present, with its corrosion consequences. The temperature and pressure are ambient, or slightly above. The density, as stated before is very close to that of the air. However, attention has to be paid to the fact that the CH_4 individually has a density of 0.72 kg/m^3, so that under segregation tends to rise. This is important when it comes to the design of the ventilation in closed areas. Explosive mixtures are formed with air, depending on the CO_2 content (Table 5).

<table>
<tr><td>

Table 4
LAMINAR FLAME
VELOCITY IN AIR (m/s)[a]

Methane	0.43
Propane	0.47
Butane	0.45
Hydrogen	3.5
Gasoline	0.5
Biogas[a]	
(55% CO_2)	0.035
(45% CO_2)	0.106
(35% CO_2)	0.179
(25% CO_2)	0.251

</td><td>

Table 5
FLAMMABILITY
RANGES OF GAS/AIR
MIXTURES (%v)

Hydrogen	4.1—75
Propane	2.1—9.4
Butane	1.86—8.41
Methane	5—15
Biogas	
55% CO_2	11—33
45% CO_2	9—27
35% CO_2	7.7—23
25% CO_2	5.9—17.6

</td></tr>
</table>

[a] Calculated with the Schuster equation

$$v = v_0\left[1 - \frac{N_2\ \% + 1.67\ CO_2\ \%}{100}\right] \qquad (2)$$

v_0: velocity of pure fuel gas.

b. Vehicular Use

For vehicular use the low energetic density of biogas makes necessary its compression, and the separation of CO_2 and H_2S. Once purified to a high CH_4 content, the gas is essentially natural gas and the same compression procedure can be followed. The cryogenic solution (2 bar and $-160°C$) is also to be considered, and lately great attention has been given to this.

Many experimental plants have organized fueling stations at ambient temperature and 200 bar to load gas cylinders aboard the vehicles; which would give them a similar autonomy as a regular gasoline tankfull. Safety regulations and procedures are the same as the ones concerning CNG.

IV. APPLICATION TO STANDARD DIESEL ENGINES

A. Engine Modifications for Dual Fuel

This chapter briefly describes the necessary modifications and devices needed to convert a standard diesel engine to dual fuel. The goal is to minimize the modifications of the engine itself and to use it like it comes from the factory. When it comes to dual fueling in the search of optimal diesel fuel substitution two possibilities exist. One, consists in a constant diesel fuel injection, varying only the gas supply to meet the power needs of the engine. This, has the eventual disadvantage of overheating the injectors, and produces a poor atomization of the liquid fuel when its contribution to the total mixture gets low. The second possibility is to maintain a constant diesel fuel/biogas ratio throughout the whole operation range. This permits the handling of the above described condition according to each particular motor. Both replacement possibilities are shown in Figure 9.

1. Stationary Engines
a. Intake System

The only addition to the intake system is the air/gas mixer, which may be called "carburetor" due to its function. According to gas feed pressure, two types may be designed: the positive pressure mixer, which consists in a direct connection to the manifold, and the

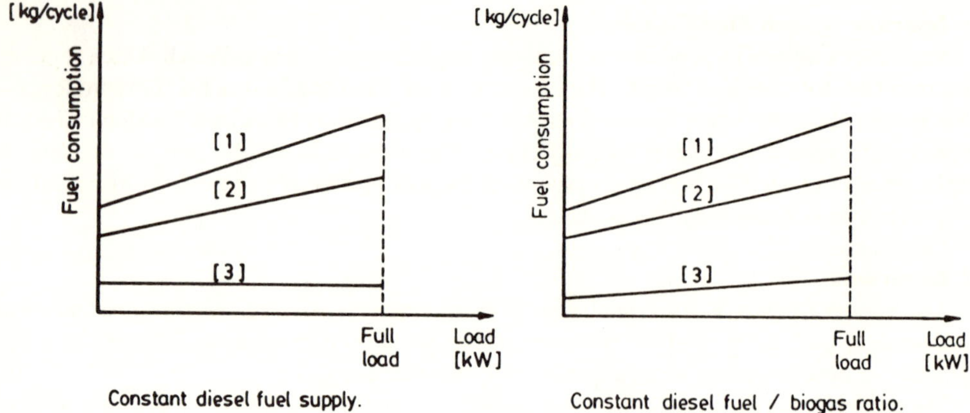

FIGURE 9.   Dual fueling schemes. Curve (1) = diesel fuel at normal operation. Curve (2) = biogas supply at dual fueling. Curve (3) = diesel fuel supply at dual fueling.

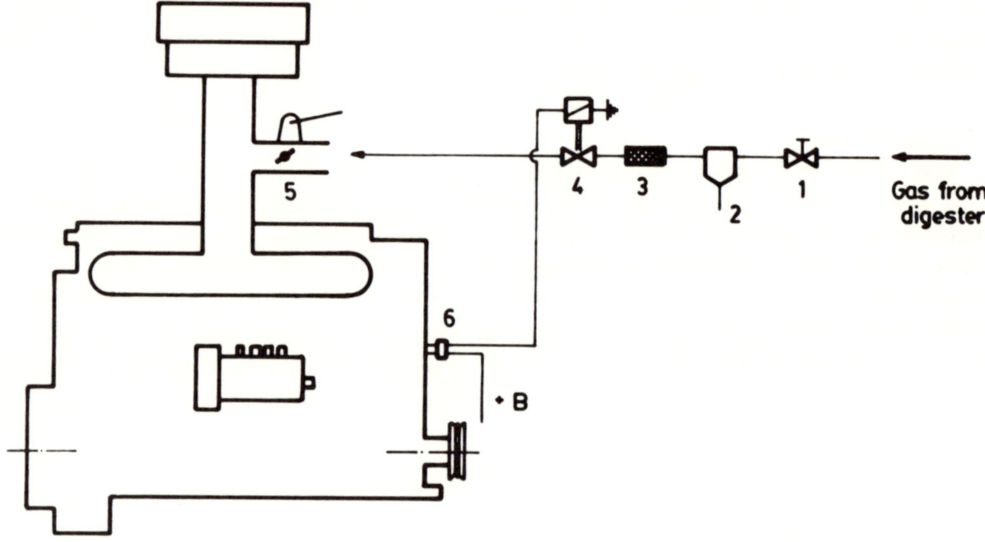

FIGURE 10.   Typical low pressure installation. (1) Cutoff valve. (2) Moisture separator. (3) Filter. (4) Magnetic valve. (5) Regulation valve. (6) Safety device.

type consisting of a venturi tube that generates a depression which is a function of the air flow. This one is necessary if biogas is supplied by a subatmospheric regulator, or by a low pressure source.

b. Gas Supply System

In most cases, stationary engines will be located near the gas source, therefore the system will be simplified to a few accessories. Among them a magnetic valve wired to a safety device is recommended. Piping size is very important for a low pressure source and must be carefully calculated for minimal pressure drop (see Figure 10).

If a subatmospheric regulator is used combined with a venturi type mixer, the system will have a safety feature: gas will not flow when the engine stops or idles. A magnetic valve is not necessary for this purpose, however a second system for safety in a diesel engine may be appropriated.

c. Injection System Modifications

Stationary engines in generator or pumping applications have a governor which regulates the injection for constant speed. Therefore no major modifications are to be introduced in this system, because when biogas is supplied the speed tends to increase, which causes the governor to reduce the diesel fuel injection. The main task will be only to regulate the replacement ratio to the maximum permitted for that particular engine. However, a diesel fuel overdose is convenient for starting.

d. Governors

According to Figure 9, there are basically two alternatives for stationary engines powering independent generators or other machines. If the diesel oil consumption is limited to a fixed quantity, a new governor will be needed only for the gas valve.

For constant gas/diesel ratio the best choice (and the most expensive) is to use a strong governor, such as centrifugal-servohydraulic or alternatively some electronic type. These types of governors can be used successfully to drive the fuel rack plus the gas valve and are not restricted to butterfly valves.

Another possibility using an inexpensive type of governor, valid for whichever substitution ratio used, consists in keeping the original one in its original function. A second governor is needed to control only the gas. However, proper performance depends on a careful matching between them, to keep the desired gas/diesel fuel ratio.

e. Safety Devices

Apart from the classical engine protections (oil pressure and water temperature), it is recommended to prevent gas from being released out to the atmosphere. A subatmospheric regulator will provide this feature. Otherwise, another device, such as an electromagnetic valve can be used. For alternators synchronized to the main supply line, it is convenient to wire the valve in such a way as to shut-off the gas supply if the generator becomes motorized or disconnected from the line. Also overspeed cero diesel oil injection must be avoided. Some special features have been developed, called "sniffers", which sense the presence of gas.

2. Automotive Engines

For better description, they can be divided in several topics, as before.

a. Intake System

The fact that not only the load but also the speed is variable, presents more difficulties to meet the conditions laid down in Section III.B.3. One method to achieve the conditions is to use a "partial flow venturi tube". (See Figure 11). Additionally, the safety specifications and the conditions of vehicular vibrations are important topics to be considered, as well as the "crush behavior".

b. Gas Supply System

Even if direct biogas is acceptable for a stationary engine, for vehicles this would increase the gas compression, bottling, storage, and transport costs. Therefore it is required to remove the CO_2. Also, for the protection of the more complex equipment used, moisture and H_2S should be eliminated.[12]

The pressure regulator needs special attention. Due to the high input pressure, it is usually designed in two stages, sometimes built in a single body. Usually, the selection of a suitable regulator to match a particular mixer, is based on the main characteristics of the engine and its use. Several manufacturers supply regulators for converting otto engines to CNG. Most of them are also suitable for diesel engines.

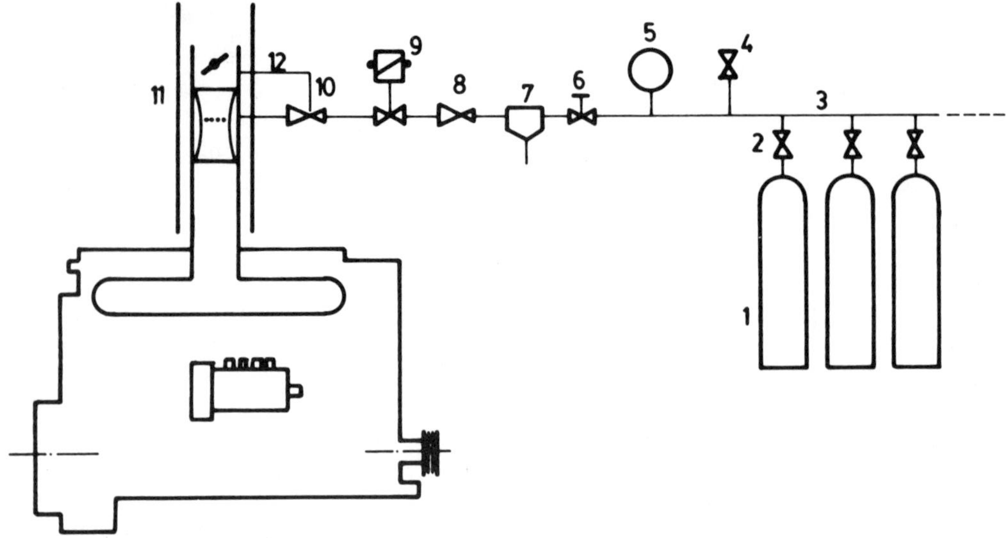

FIGURE 11.   Typical vehicle installation. (1) High pressure storage bottles. (2) Individual bottle valves. (3) Supply manifold. (4) Charging valve and connection. (5) Pressure gauge. (6) Manual shut off valve. (7) Strainer and moisture separator. (8) First stage regulator. (9) Magnetic valve. (10) Second stage regulator (Sub-atm). (11) Venturi type mixer. (12) Pressure reference line.

The other components of an installation are: tubes, fittings, pressure gauge, charging valve and connection, cut-off valves (magnetic and manual), and control and safety devices. Regulations and standards concerning gas piping and fittings should be followed.

c. Governors and Safety Systems

The different uses given to the vehicular applications will be responsible for the necessary changes in the governors and safety systems as explained before. As an example, the modifications described in Section V.C, may serve as a guide for specific cases.

B. Performance

The results obtained in the research program developed at the UTFSM may serve as an average behavior for standard direct injection diesel engines using biogas and torch-ignition (dual-fuel).[13]

Biogas from a sanitary landfill was used and the engine was a MWM (D-229-4/42 kW/1800 RPM) of Brasilian make. The stroke was 120 mm and the cylinder diameter 102 mm. The compression ratio was 16.6:1. Figure 12 shows the test bench installation used.

1. Full Load Behavior

At full load the data of the makers for diesel oil running were used. The brake specific fuel consumption (BSFC) equivalent to diesel oil were obtained, as shown in Figure 13: Increasing the energetic proportion of biogas the BSFC grows showing a worse performance on higher revolutions.

In Figure 14 the exhaust gas temperature and some components variations are shown for two revolution ranges and the biogas energetic replacement ratio ψ. The exhaust gas temperature has a falling tendency when the biogas replacement ratio λ is increased, with differences up to 50°C. This situation may be understood if the increasing inert gas consumption (CO_2) is considered, even if the suctioned air is reduced (assuming constant volumetric efficiency). When reduced air is introduced, the air ratio does not suffer great variations, since the stochiometric air decreases with more biogas provision. The emission

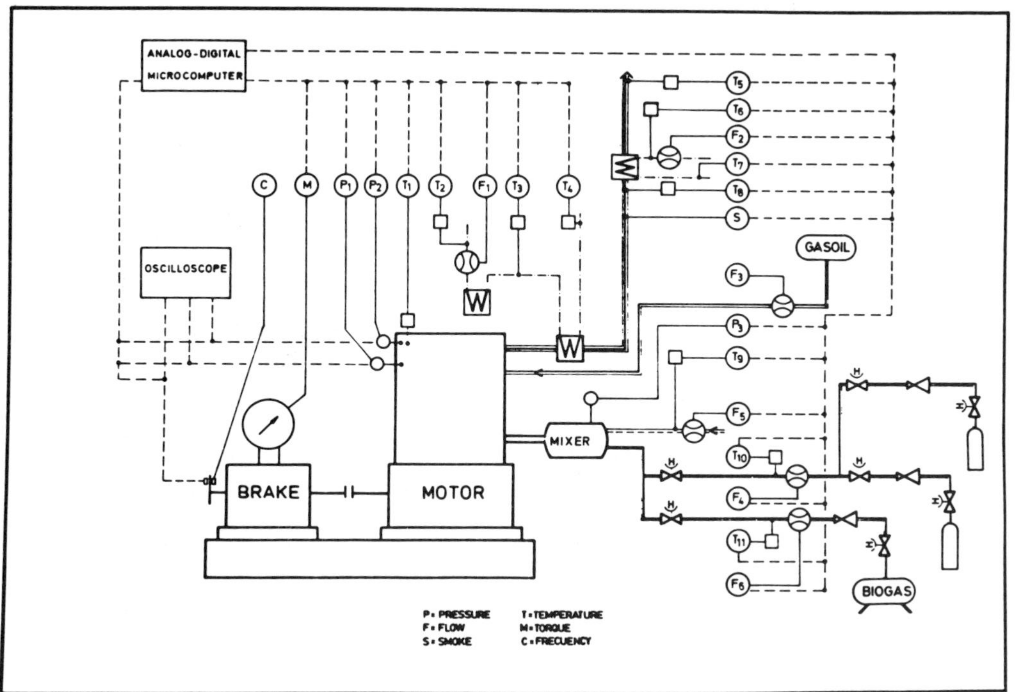

FIGURE 12.　Test Bench at the UTFSM Laboratories.

of soot is highly reduced (1/3 to 1/4 of diesel oil only) with the biogas supply ratios experimented. But the CO concentration doubles at the same conditions. The falling O_2 concentration may be due to the lesser sucked air.

2. Influence of the Biogas Composition, Full Load

Comparisons with 56% CH_4 biogas and pure methane were made. As reference points the only diesel oil runs are given in Figure 15. The thermal efficiency expressed in BSFC, is very similar in both extreme conditions (only diesel oil and pure methane), using 80% of biogas (and CH_4) energetic supply. The measured intermediate values show less efficiency with biogas containing more CO_2. Most probably this is due to the inert gas (CO_2) presence. Both reported revolutions show a similar development of the BSFC (equivalent) as function of the biogas methane content.

The exhaust gas temperatures decrease with a high CO_2 content in the biogas but reducing the CO_2 until pure methane similar levels as with pure diesel oil operation are reached. The CO_2 reduction using pure methane is explained through the C/H ratio differences with only diesel fuel, and the high CO_2 contents with biogas is due to the CO_2 addition with this gas in the feed.

High CO concentrations were measured, showing two and four times as high as only diesel oil running. Even higher values were encountered with biogas (56% CH_4) which may be due to the lower temperature levels and the lower flame velocities.

The soot values obtained are greatly reduced with methane ($\psi = 80\%$) in comparison to only diesel oil operation, and still more reduced for CH_4/CO_2 mixtures. This may be because of the CO_2 of the biogas creating higher turbulence in the combustion chamber. Also the methane/total reactants ratio is lower for the biogas (56% CH_4) as for pure methane, conditioning probably a different mechanism in the soot formation scheme.

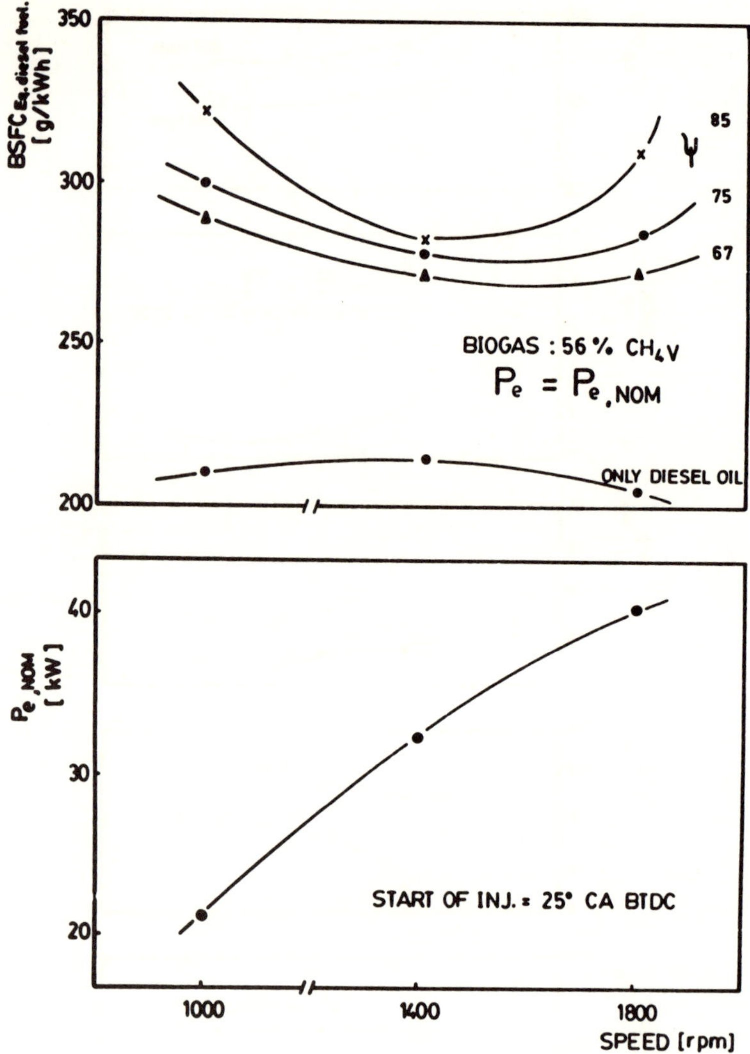

FIGURE 13.   Effective power and brake specific fuel consumption (BSFC).

3. Influence of the Injection Point, Full Load

To study this influence, each experimental series was begun with the nominal injection point (25 BTDC), measuring the relevant parameters as a function of the injection point variation. The power differences due to the biogas supply ratio change (ψ = 75% and ψ = 67%) are insignificant, so only one curve was drawn for the results obtained (Figure 16).

The power decay when retarding the injection point shows a higher sensibility in the upper revolutions range as compared to the intermediate range. But the sensibility of power decay when the injection point is advanced, is greater with only diesel operation. The BSFC reflects the commented behavior and the minimum values correspond to the maximal power obtained.

The exhaust gas performance can be seen through the soot numbers and the CO content in Figure 17. The influence of the injection point on this components is lower for biogas operation than for only diesel operation, even if the absolute values for this condition lie on different levels.

4. Cylinder Pressure Development

The mechanical stresses were evaluated through the cylinder pressure oscillogram. For

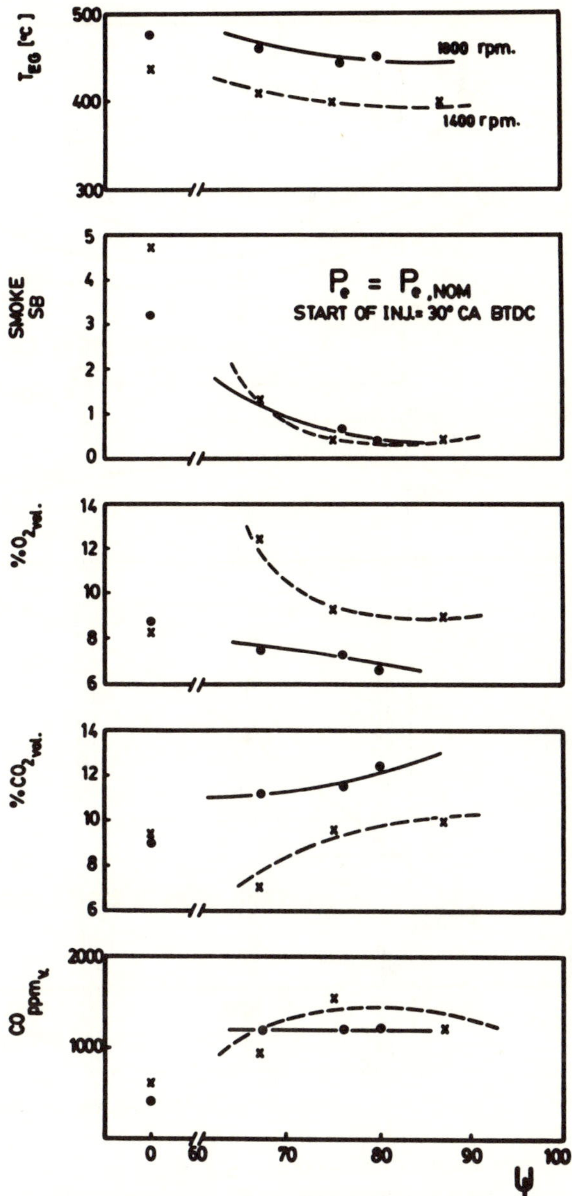

FIGURE 14. Exhaust gas parameters.

full load (Pe = Pe,NOM) in Figure 18 the peak pressures as well as the maximal pressure gradients as a function of ψ are seen. The peaks shown correspond to the mean value of them, but the ripple grade grows with the biogas supply.

As an example in Figure 19 the sequence of the peaks are shown. In the range of 65 to 85% of ψ, the peaks tend to decrease when ψ is augmented, but if ψ grows up to 95% this peak will be higher than the value for diesel oil only. Similar are the curves for the maximal pressure-gradient. During the experimental runs no detonation was detected, as long as the energetic supply was limited only to obtain the nominal power. As soon as these supplies were increased, detonation noises and abnormal pressure-developments were found.

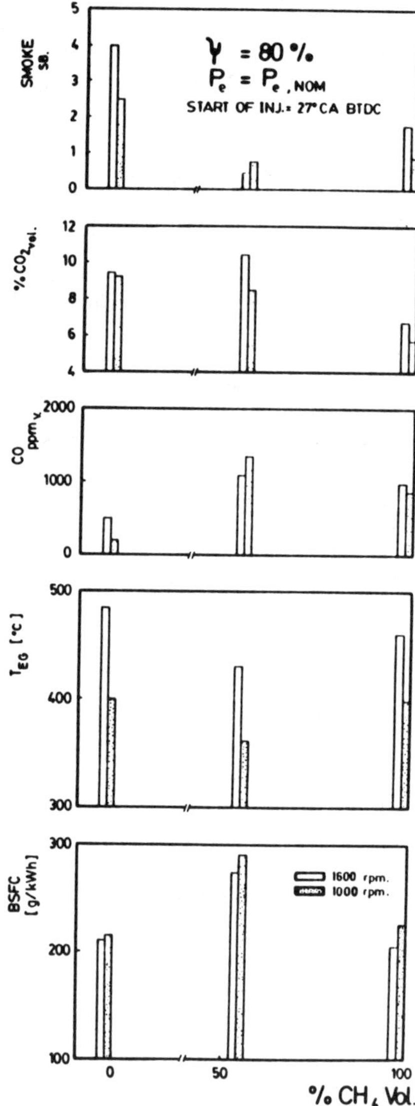

FIGURE 15.   Comparative performance at different substitutions and RPM.

5. Summary of Performances Described

The sulfur content in the biogas used presents no problems in the motor itself, but it may cause corrosion in heat exchangers if TOTEM operation is intended. Soot emission is greatly reduced with this operation, however increasing the CO emission; but the values reached are below the maximum permitted. Better performance can be obtained with adequate injection point modifications; however it may fall with incorrect procedures. Also this dependency can be generally stated when air/fuel ratios are modified. Thermal efficiencies and power output of engines will vary with biogas quality (% CH_4) and with the energetic replacement ratio (ψ).

No serious difficulties were found when using standard diesel motors with biogas when reasonable replacements and power outputs were applied. Only high biogas supply causes abnormal function. Referring to the combustion chamber design, the different temperature

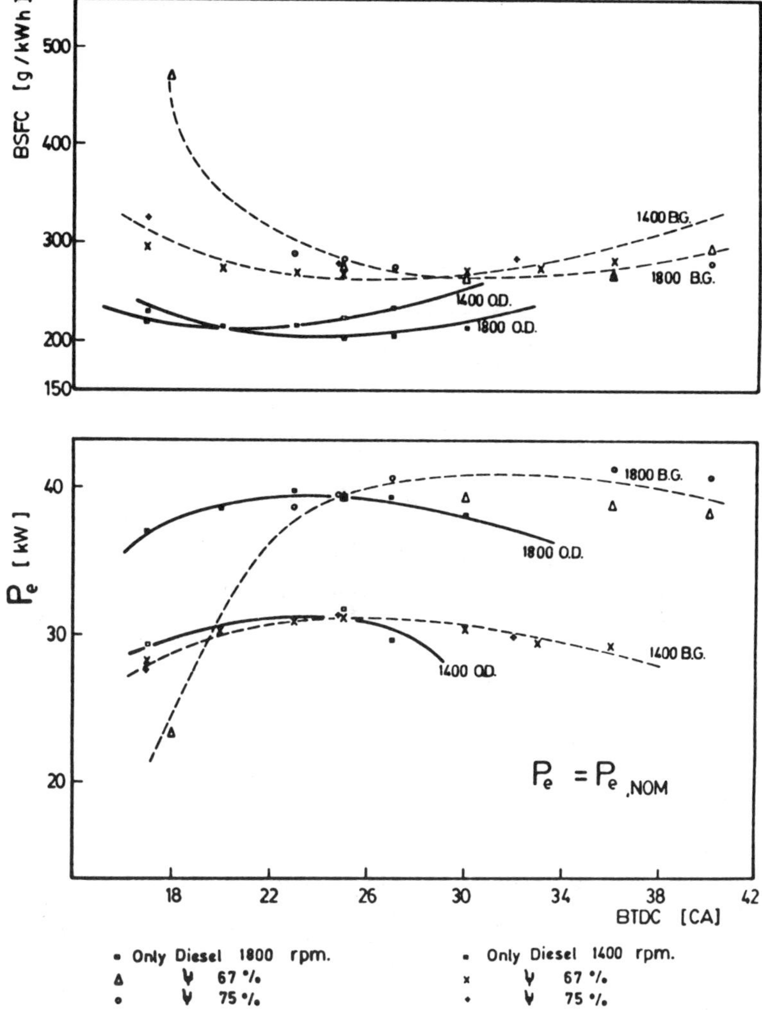

FIGURE 16.   Power and BSFC vs. start of diesel fuel injection.

gradients and combustion developments can conduct to hot points or injector clogging, magnified by the reduced cooling effect due to the lesser injected diesel fuel. In conclusion, each type of standard motor should be thoroughly test run before being declared adequate for dual-fuel operation.

V. PRACTICAL EXPERIENCES

A. Illapel Project

1. Introduction

Illapel is the name of a small city about 250 km north of Santiago. In 1982 an industrial biodigester designed by BIMA (Austria) was installed to receive the manure of a private pig farm. The farm counts 3000 pigs which produce about 7 tons/day of fresh manure. The maximum capacity of the digester is estimated for 4000 pigs. A floating reservoir is capable of storing 250 m³ of gas above the effluent pond.

The digester is kept at 35°C constantly by means of a gas fired boiler and circulating water. The gas production is about 320 m³/day when operated on the manure of 1300 pigs

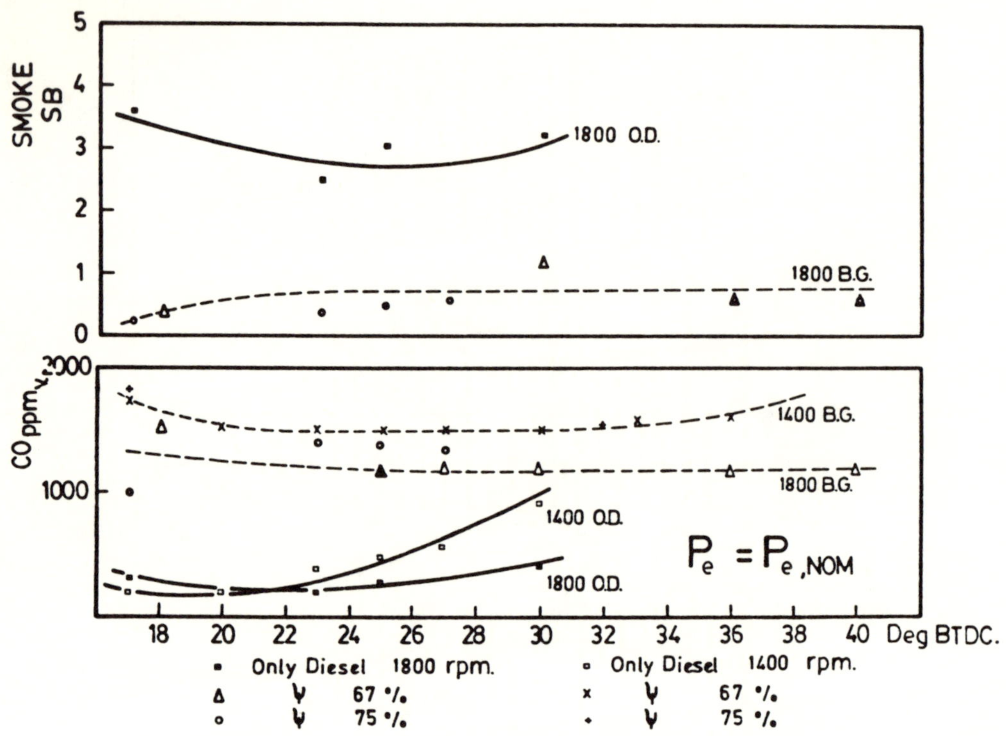

FIGURE 17.   Smoke and CO vs. start of diesel fuel injection.

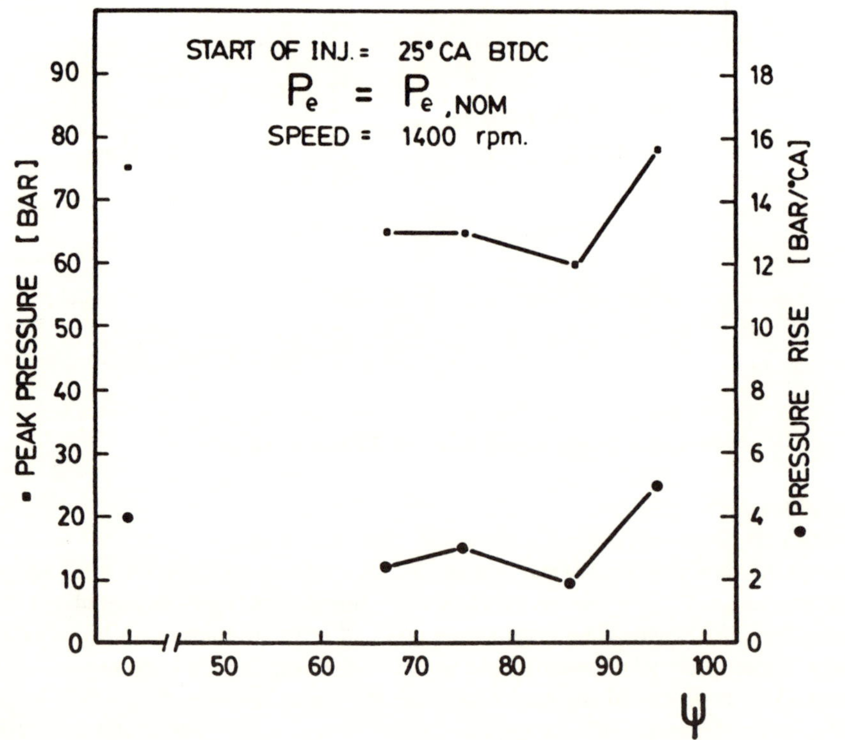

FIGURE 18.   Peak pressure and pressure rise in the combustion chamber.

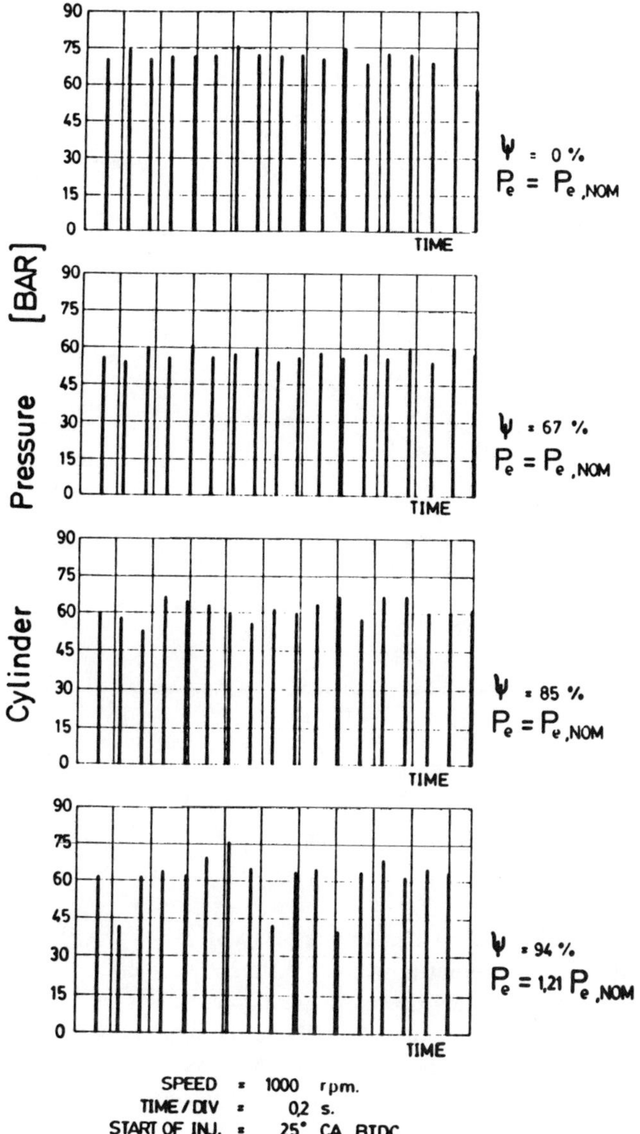

FIGURE 19.   Cylinder pressure oscillogram peaks.

and the water heater uses 20 m³/hr. Its burning time depends on the season and ranges from 8 to 12 hr. That leaves a net gas availability of 80 to 160 m³/day.

2. Installation

Our research group performed a series of test runs with the purpose of studying the operation of the motor/generator and digester system.[14] The set is composed of a MAN 9614 diesel engine (4 stroke, 4 cylinders) and an AC generator (AEG, 3 phase, 50 Hz, 380/220 V) developing 23 kW continuous at 1500 rpm. The engine was fitted with two heat exchangers to recover waste heat from cooling system and exhaust gases. Several experiments were performed with variable load and gas/diesel ratio. For the first part of the experimental work, a 20 kW 3 phase resistor box was used as load.

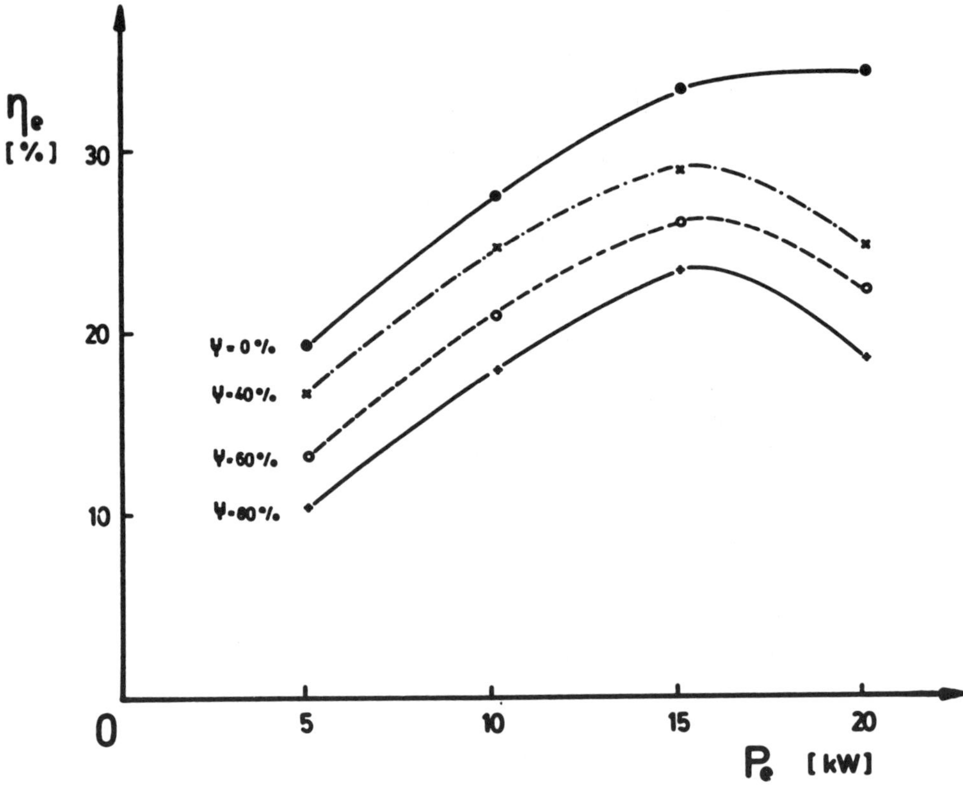

FIGURE 20.   Engine efficiency vs. output power.

3. Results

Results of these runs are shown in Figures 20 and 21. Figure 20 represents the engine efficiency as a function of output power for different energetic replacement ratios ψ. It is noticeable that thermal efficiency decreases for higher values of ψ.

It is interesting to know the value of available thermal energy if the engine is installed to work under the total energy concept (TOTEM). As it can be seen in Figure 21, this thermal energy increases with output power and gas replacement ratio, as a direct consequence of efficiency variations.

a. Long Term Operation Test

The second part of the experiment consisted in supplying the farm directly with electrical energy produced by the generating set. It was done every day for several weeks, at the time of the food mill operation (from 4 to 6 hr/day at an average power of 15 to 19 kW). The gas replacement ratio was kept as high as possible, around 80 to 90%. Oil samples were taken after 24 and 48 working hours for analysis, and degradation was found to be less than expected for only diesel operation.

b. Fuel Cost Analysis

The mill uses 10 to 15 kW for 4 to 6 hr about 20 times/month. The rest of the farm needs about 0.5 kW during daytime and near 5 kW at night. The average monthly bill is $300 (U.S. equivalence) for 2500 kWh. Fare application is complicated, depending on season, type of installation, fixed charges, etc.

Every milling session, the generator produced from 60 to 100 kWh at an average power

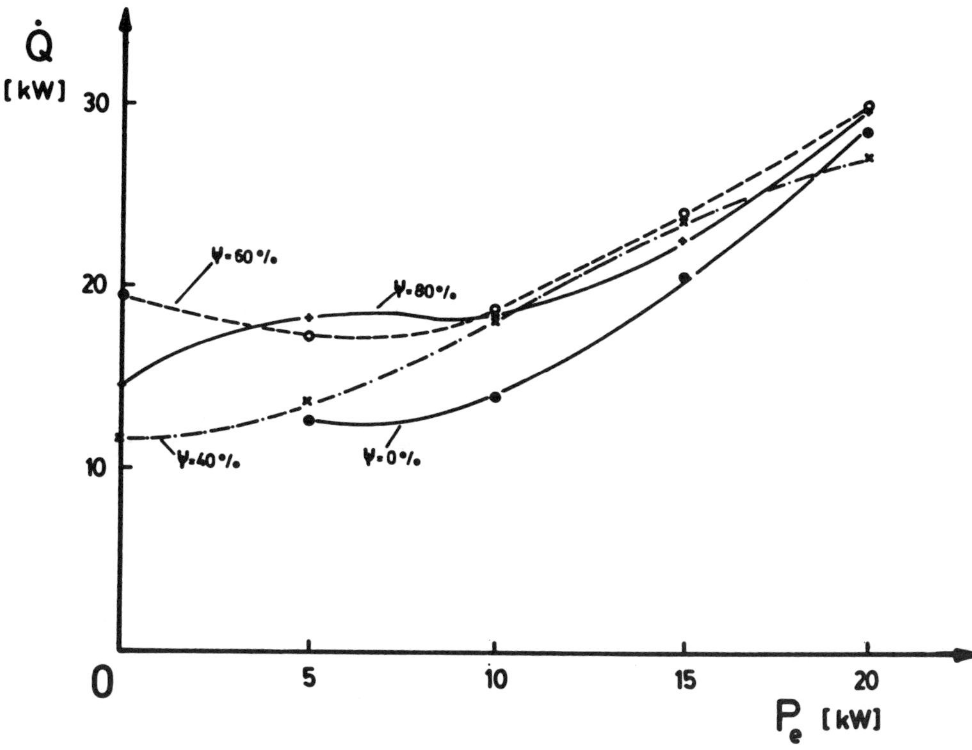

FIGURE 21. Available thermal energy for TOTEM operation.

of 15 to 19 kW (the complete farm was connected at nighttime). The average consumption of diesel fuel each time was 6 ℓ, which adds to 120 ℓ/month. The current fuel price at the time was $0.40/ℓ which means $48/month.

Electricity consumption dropped to about 1000 kWh and $120. That gives a net saving of $132/month. Maintenance cost is not considered because more long term data is necessary for a reliable evaluation. These results are represented in Figure 22.

In this particular case, gas value was set to zero, because no alternative use was made of it. The digester was built by sanitary reasons only and most of the gas was normally wasted.

B. Thermophilic Digester/Diesel Engine, Total-Energy Pilot Plant

A pilot plant of a thermaly integrated system under the total-energy principle was built at a pig farm located in Limache, some 50 km of Valparíso, Chile where the UTFSM campus is located.[15] The construction was supported by the Chilean National Fund for Scientific and Technological Development and the Technical University Federico Santa María Research Committee. It was initiated in August 1983 and put into operation in mid-1984.

A total-energy system is one which allows the use of the residual energies of any transformation process. This has found an excellent application in internal combustion engines and has led to the development of the total-energy motor known as TOTEM, already in commercial application. The main objective pursued by this project was to check the operational feasibility of the combined system at a pilot plant scale. The design requirements, based on some extent on the equipment and funds availability, were the following:

- Thermophilic operation of a 10 m³ biodigester, on a daily loading basis.
- Thermal interconnection of the digester with a 31 hp diesel engine/generator, to use its waste heats for the digestion heating requirements.

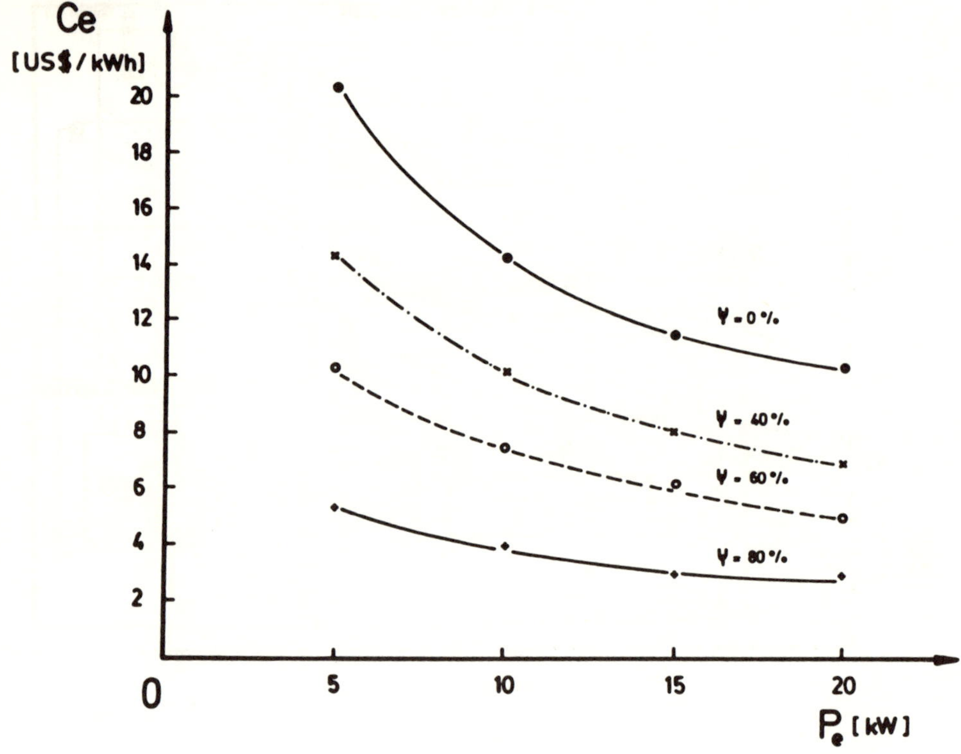

FIGURE 22.   Fuel cost as function of output power, for different gas replacement ratios.

- Ample flexibility in operation, including difference in solids and hydraulic retention times.

The following operational conditions of the biodigestion were studied:

- Total solids (TS); 4 to 8%.
- Hydraulic retention time (HRT); 20 days to washout.
- Temperature (T); 30 to 60°C.
- Solids retention time (SRT); equal or different to HRT.
- Agitation; occasional.

Simultaneously, series of test runs of the diesel engine are being performed. These were planned to understand not only the interconnection features but also to study some motor operation modes. The parameters to study are biogas substitution, air-fuel ratio, fuel injection point, and heat recovery on TOTEM operation. The final objective pursued will be to optimize the whole combined system in the scope of economy and maximal energy output.

1. Plant

The system is comprized by three main subsystems (fermentation, motor/generator, thermal control) and an operational control and data acquisition system. A simplified scheme is shown in Figure 23.

a. Fermentation Subsystem

This includes the digester with its agitator and heat exchanger, the loading tank with its

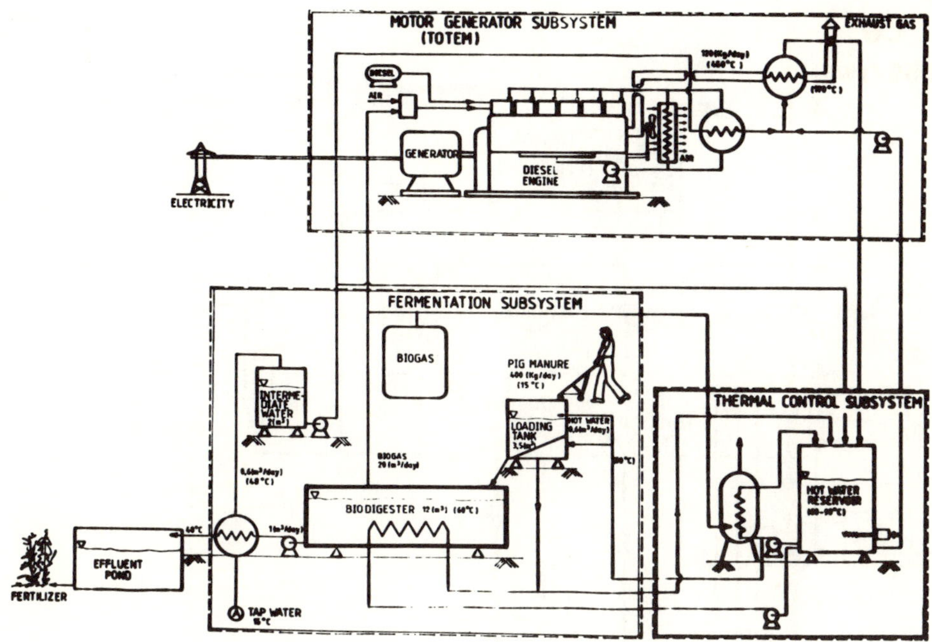

FIGURE 23.   Flow diagram of pilot plant.

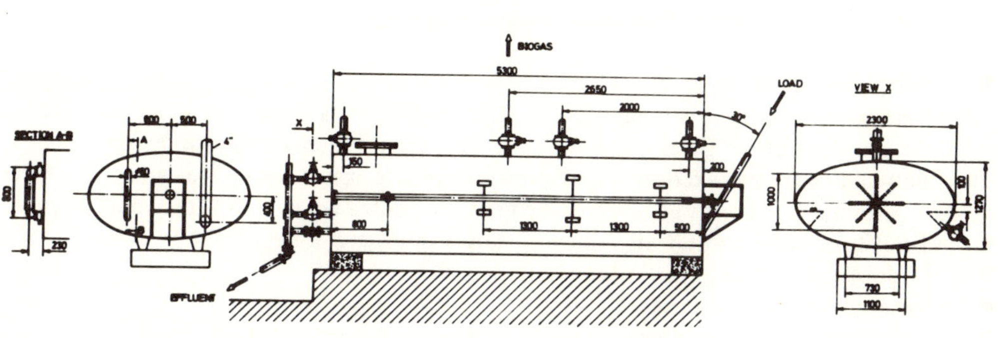

FIGURE 24.   Diagram of 12 m³ digester constructed from a truck tank.

own heat exchanger, and the effluent heat exchanger and receiving pond. The gas storage is also included in this subsystem.

The *digester* was made from an old molasses truck tank, (cut down to 12 m³) with a mixer of four two-bladed axial turbines. The mixer is driven by a 1.5 hp monophasic electric motor through a reduction gearbox, at 12.8 to 30 rpm. A detail is shown in Figure 24.

The *loading tank* is a 3.5 m³ mechanically mixed rectangular vessel with a double bottom as heat exchanger. The prepared load is discharged into the digester by gravity through a 4 in. pipe, designed to avoid air entrainment.

The *effluent* is discharged to an open cement pond from where it is channeled to the downhill fields for agricultural use.

The *gas* is stored at 0.02 bar in a 25 m³ reinforced plastic bag after being measured, and after passing drip and flame traps.

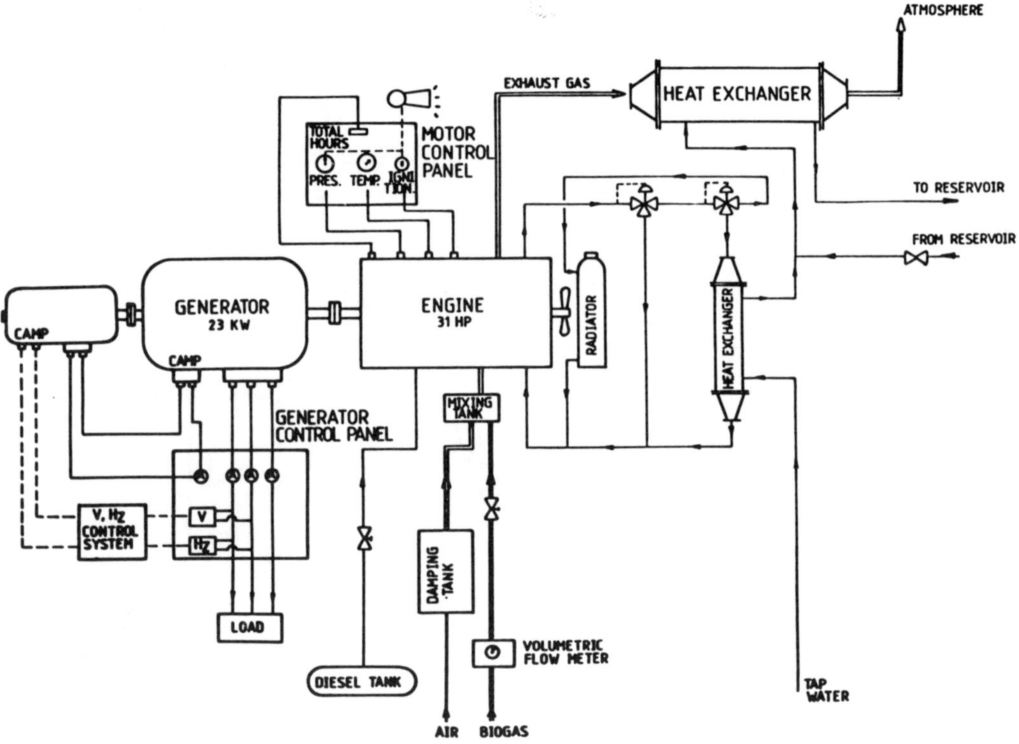

FIGURE 25.   Motor/generator system.

b. Motor/Generator Subsystem

The test-bench available is comprized by a 31 hp MAN standard diesel engine, coupled to a 23 kVA three phase generator with external exciter and electronic frequency control. Two shell and tube heat exchangers recover the cooling water waste heat and part of the exhaust gases thermal energy. The gases are not to be cooled below 170°C to avoid corrosion problems. The scheme of this subsystem is shown in Figure 25.

c. Thermal Control Subsystem

Since the daily biogas production is only enough to operate the diesel engine for up to 2 hr, a heat reservoir had to be incorporated to store all the recoverable waste heat from the motor for its use when needed by the fermentation subsystem. This in no way invalidates the interconnection results to be obtained, because the thermal control is inherently non-continuous. The major heat requirements are to heat up to the operating temperature when the daily load is prepared.

Only occasionally (2 to 4 times a day) does the internal heat exchanger operate within the digester to restore the design temperature (within the preset ± 0.5°C); because of minimal heat losses to the surroundings. The motor is not run every day, so that a biogas fired water heater is incorporated in this subsystem.

d. Operational Control and Data Acquisition Subsystem

This consists of a series of local single loop: level, flow, and pressure controllers of the various interconnected equipment. A central microcomputer was planned to be installed during the second year of operation of the plant, for the data acquisition and operational control.

Table 6
SYSTEM MASS BALANCE

Item	Flow kg/hr (ℓ/min)	Quantity kg/day (m³/day)	Total time of operation min/day
Fresh pigsty solids (20% TS)		400	Wheelbarrow loading = 80
Warm water to loading tank	(50)	600	Mixing of load = 12
Digester loading (8% TS)	(250)	1000	Loading process = 4
Digester effluent	(90)	1000	Discharge process = 11
Warm water to digester's internal heat exchanger	(50)	2100	Supplement loading temp and recover heat losses = 42
Fresh water to effluent			

The whole installation is at a 1000 units pig farm of Thomsen & Co. (Chile). The facility is constructed on a 10% slope hillside so that gravity flow of most of the liquids is possible. The pilot plant is installed in a 100 m² building which also covers the plastic gas-bag to avoid UV degradation. The total cost of the plant was $25,800 — with a real value of $57,800. The difference is due to the ample utilization of scrap materials, considering a shorter life expectancy of the pilot plant with respect to an industrial installation.

In the former figures the value of the fully instrumented diesel engine/generator ($27,000) has not been considered because this equipment is on a loan basis from the Thermofluids Labs of the UTFSM.

2. Mass and Energy Balances

For design mass and energy balances the operational conditions assumed were

Temperature, (T) = 60°C
Total solids, (TS) = 8%
Volatile solids, (VS) = 6%
Hydraulic retention time (HRT) = 10 days
System pressure (P) = 0.02 bar
Continuity of operation = once a day loading
Digester liquid volume = 10 m³
Gas storage capacity = 25 m³
Volatile solids degradation (DG) = 42%
Productivity: gas volume (VG) = 20 m³/day; conversion (ED) = 0.800 m³/kg VS degraded; and gas density (ρ) = 1.15 kg/m³.
Heating values: gas (HBG) = 20150 [kJ/m³]; diesel (HD) = 44000 [kJ/kg].

The mass balance at the above conditions is shown in Table 6. The corresponding energy balance is shown in the Sankey diagram included as Figure 26. The mass and volumetric flows and temperatures of the different currents are shown in Figure 23, and a photograph of the installation as Figure 27.

C. Municipal Garbage Collection Truck

In the first semester of 1984 a joint project between the Municipality of Santiago and the Technical University Federico Santa María was developed. The aim of this project was to use the biogas of the "La Feria" sanitary landfill, as fuel for a garbage collection truck. The technical feasibility should be proved and the variation of the ambient contamination, measured as Bosch Soot Number, had to be evaluated.[16]

The biogas had to be purified, for which a rudimentary $Ca(OH)_2$ absorption tower was

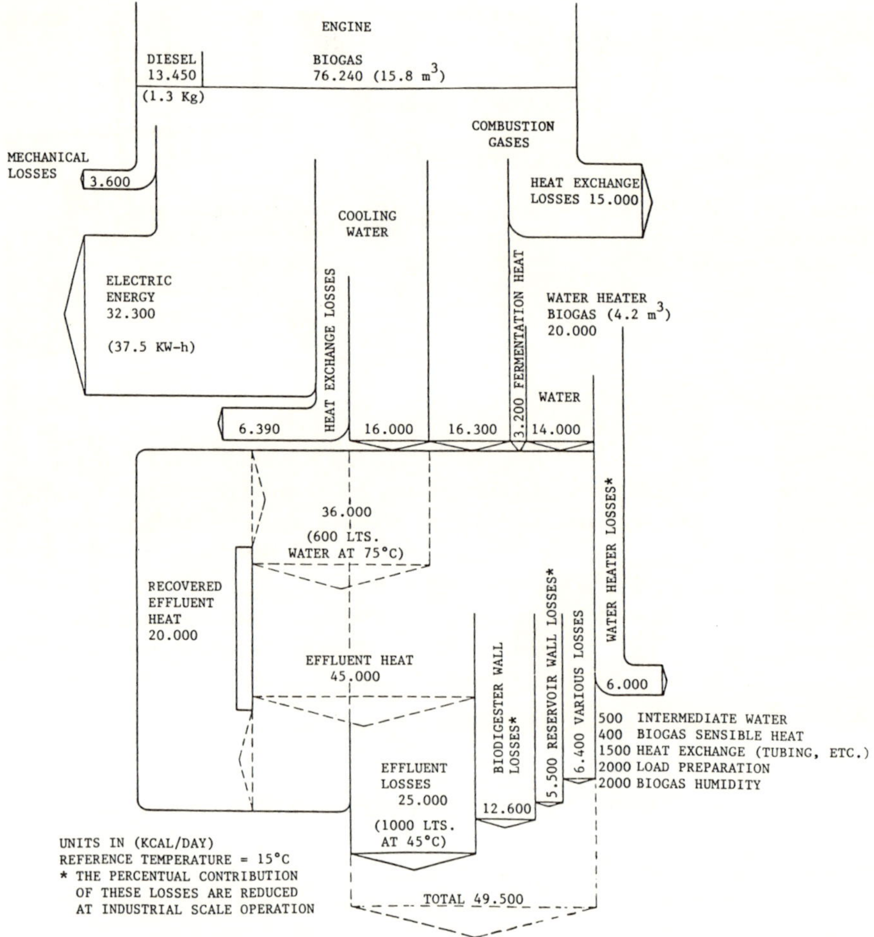

FIGURE 26.   Sankey diagram of pilot plant at the design operation point.

constructed. The same power output as reached with diesel fuel only was desired and the interchangeability between diesel fuel and dual fuel operation with only a selector key was to be obtained. Diesel fuel consumption reduction was reached and also a significantly less ambient contamination due to soot in the exhaust gases.

1. Biogas Cylinders

As seen in Figure 28, the autonomy range is a function of the total volume and pressure of the cylinders used. The limited resources, the main goals and the technical possibilities available led the last decision to be, 10 (0.1 m³) cylinders at 20 bar.

2. Motor Modifications in Test-Bench

First a theoretical approach to the desired dual fuel supplies was made, which had to be proved experimentally. To find out the correct modifications, the engine was mounted on a test-bench at the UTFSM where the real work conditions could be simulated.

Different systems of dual fueling were tested, and the Venturi in an anular section space with counter butterflies was optimized for the desired conditions, (ψ = 80%). Figure 29 shows this gas supply feature. To avoid malfunction and explosion danger the diesel fuel injection system was modified to withhold a cero-injection situation during deceleration and downhill traveling. When the engine stops and when over-revolution may occur the biogas supply must be cut-off. To take care of this, the arrangement shown in Figure 30 was tested and then installed.

FIGURE 27.　Pilot plant in operation at pig farm.

3. Performance in Test-Bench

In Figure 31 the full load curve obtained is shown. The supplied power relation (ψ) practically is held constant.

4. System Installation in Truck

Schematically the installation is shown in Figure 32. The butterfly position is directly linked to the operation pedal and coupled to the injection pump.

5. Performance in City

Figure 33 shows the truck in its function, under dual-fuel operation.

6. Reduction in Contamination

In Figure 34 the soot reduction is shown, according to the Bosch number. As can be seen, the reduction obtained corresponds to 85%.

7. Costs

The costs of the materials for the modifications gave the sum of $1841, and detail is shown in Table 7.

8. Conclusions

The previous laboratory tests permitted the correct modifications to the supply systems of biogas, air, and diesel fuel. At full load, up to 90% diesel oil replacement was used but these high biogas contents are not recommended for this motor, because of the high solicitations.

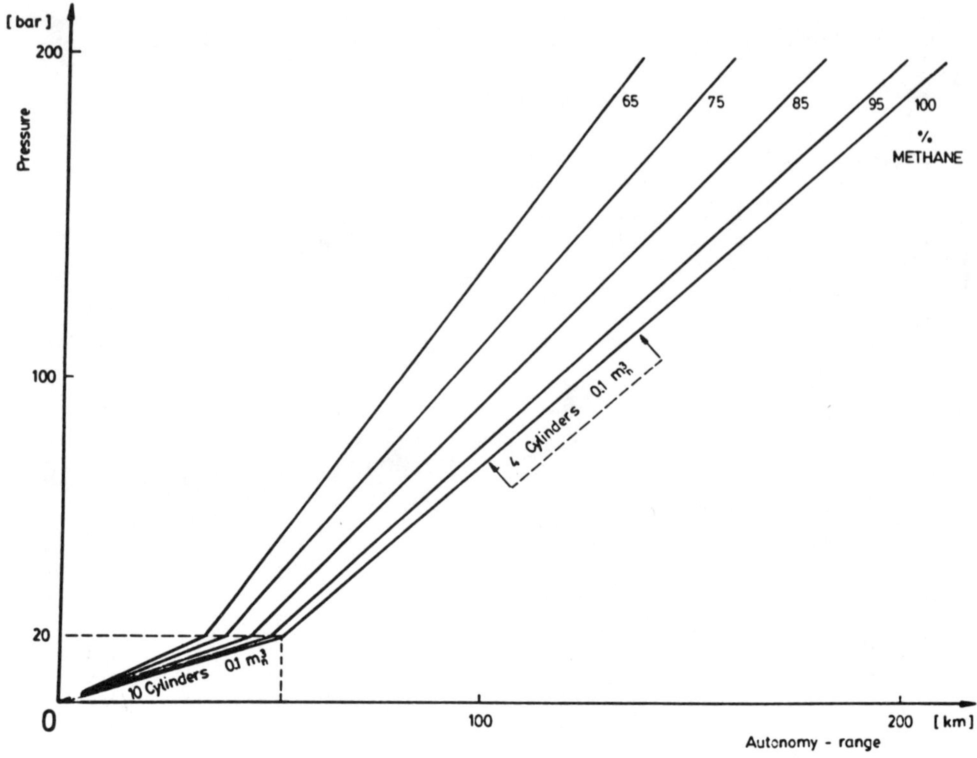

FIGURE 28.   Truck autonomy range for different gas cylinder pressures.

- The simple gas treatment tower constructed was capable to produce biogas with less than 5% CO_2, ready for compression.
- The system designed worked satisfactorily. It also readily permitted the conversion of dual operation to only diesel while the truck was running.
- The soot contamination was reduced to only 14% of the only diesel functioning.
- The truck autonomy with this dual-fuel system (20 bar) and 70% energetic replacement was 48 km, which may reach 200 km, using a set of 4, 200 bar cylinders with 0.1 m^3 compressed gas.

VI. NOMENCLATURE AND ABBREVIATIONS

A/F	— Air to fuel
BSFC	— Brake specific fuel consumption (g/kW · h)
BTDC	— Before top dead center (° crank shaft angle)
C/H	— Carbon to hydrogen
Ce	— Diesel fuel costs (U.S.$/kWh)
CA	— Degree camshaft angle
CN	— Cetane number
CNG	— Compressed natural gas
DG	— Volatile solids degradation (%)
ED	— Conversion (m^3/kg) VS degraded
HBG	— Biogas heat of combustion (kJ/kg) or (kJ/m^3)
HBL	— Humid biomass loaded
HD	— Diesel fuel heat of combustion (kJ/kg)

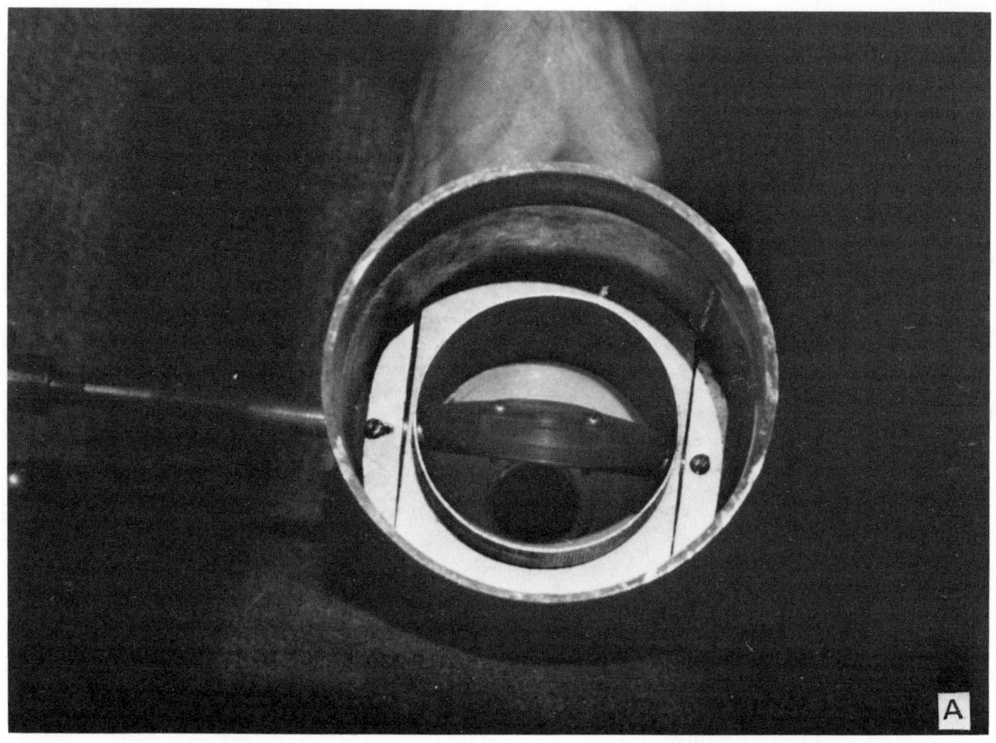

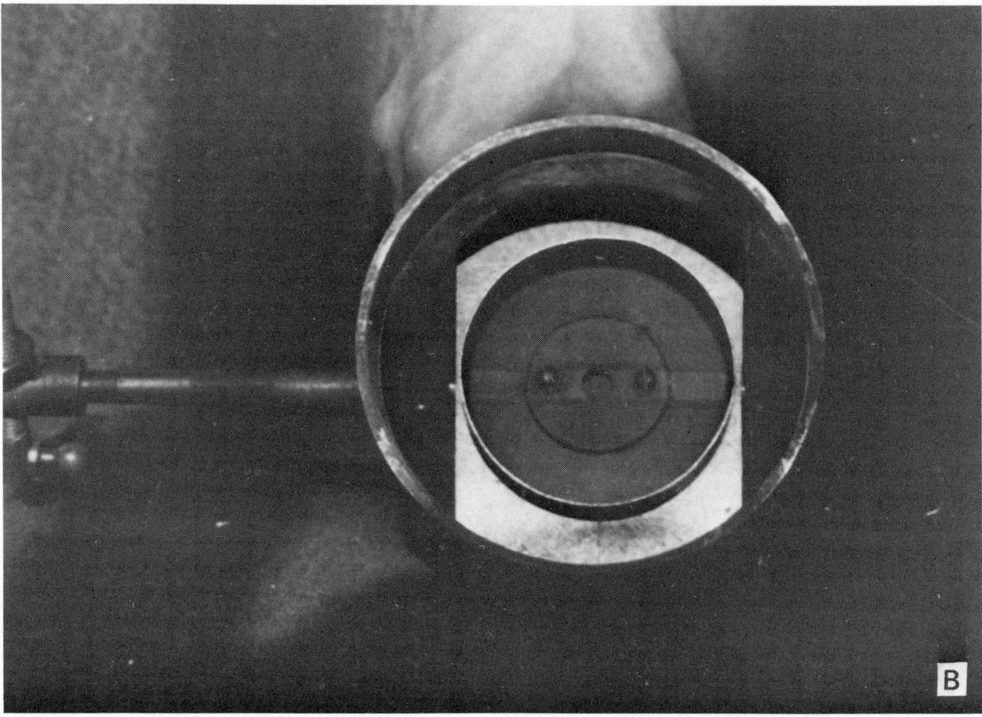

FIGURE 29.   Double butterfly for gas supply.

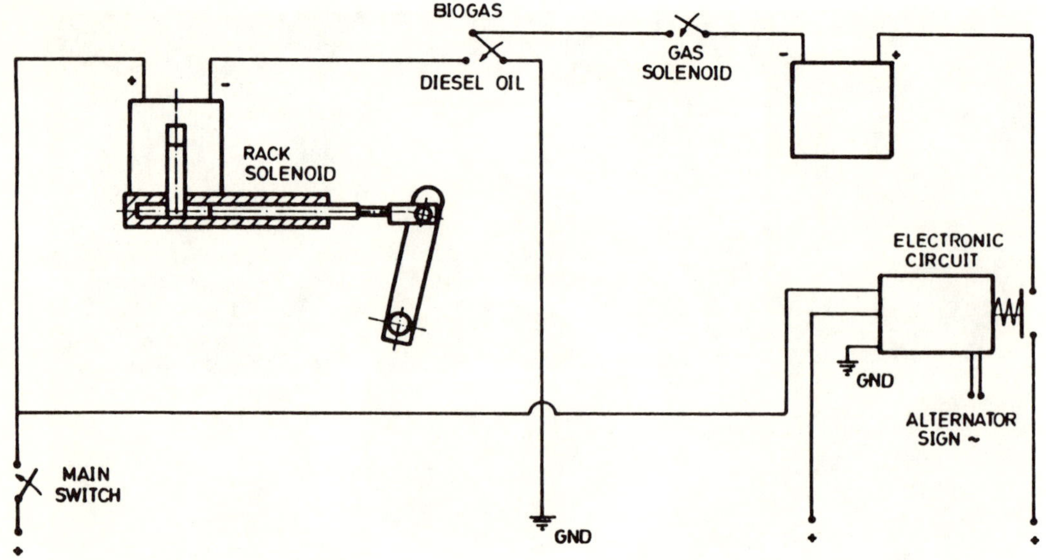

FIGURE 30.   Solenoids connection scheme.

HRT	—	Hydraulic retention time (days)
Hi	—	Heat of combustion
LPG	—	Liquefied petroleum gas
M	—	Torque (N · m)
mfBG	—	Mass flow of biogas (kg/h)
mfD	—	Mass flow of diesel fuel (kg/h)
n	—	Speed (rpm)
ON	—	Octane number
P or p	—	Pressure (bar)
Pe	—	Effective power output (kW)
Pe, NOM	—	Nominal effective power output (kW)
Q	—	Available heat flux (kW)
SB	—	Soot Bosch
SRT	—	Solids retention time (days)
T	—	Temperature (°C)
TOTEM	—	Total-energy motor
TS	—	Total solids (%)
UTFSM	—	Universidad Técnica Federico Santa María (Technical University Federico Santa María, Valparaíso-Chile)
v	—	Laminar flame velocity (m/s)
v_o	—	Flame velocity of pure fuel gas (m/s)
VG	—	Gas volume (m³/day)
VS	—	Volatile solids (%)
η_e	—	Effective efficiency
λ	—	Actual air/stoichiometric air, ratio
ρ	—	Density (kg/m³)
ψ	—	Energetic replacement ratio (%)

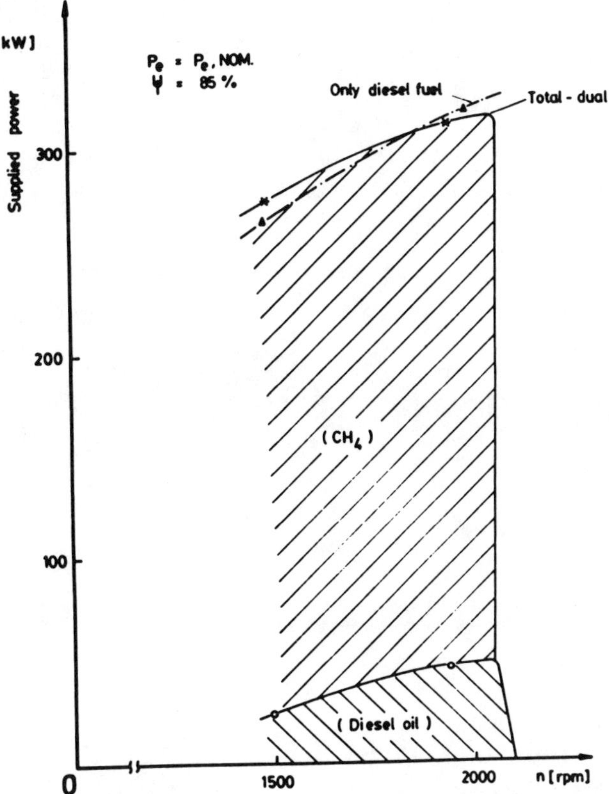

FIGURE 31. Supplied power for nominal output.

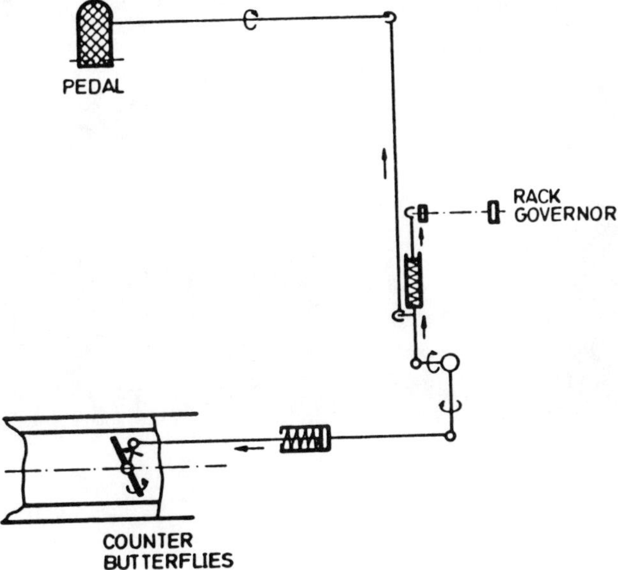

FIGURE 32. Linkage installation in truck.

FIGURE 33.   Truck in operation.

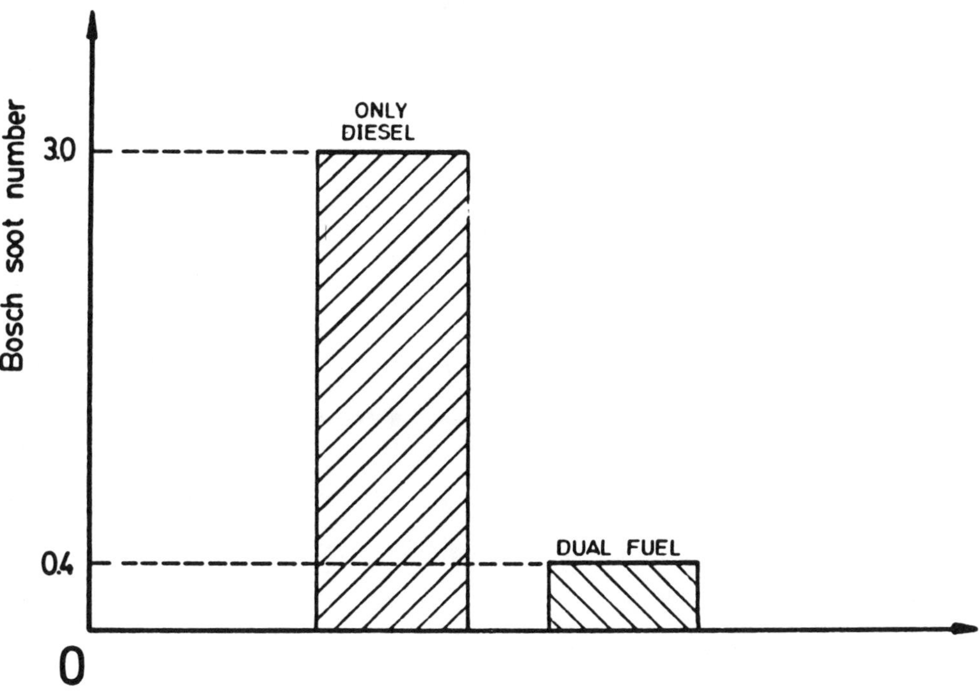

FIGURE 34.   Soot Bosch number reduction.

Table 7
COST OF MATERIALS FOR ENGINE
MODIFICATION TO DUAL FUEL

Primary pressure regulator	$280
Secondary pressure regulator	$200
Solenoid valve	$ 60
Electronic circuit (acting on solenoid valve for biogas cutoff)	$150
Manometers	$100
Gas cylinders	$471
Cylinders installation	$480
Counter butterflies system	$100
TOTAL (June 84)	$1,841.

REFERENCES

1. **Wong, J. K.,** Study of mixtures of methane and carbon dioxide as fuels in a single cylinder engine (CLR), SAE Paper No. 770796.
2. **Alkalaj, D. L., Sáez, A. C., and Moltedo, C. M.,** La combinación Biodigestor-Motor-Generador como Sistema de Energía Total, paper presented at 3rd Conference on Solar, Wind and Related Energies, Valparaíso, Chile, November 9 to 12, 1982.
3. *Diesel-LPG,* Landi den Hartog bulletin on Diesel/LPG Dual Fuel System, Vienna, 1984.
4. **Ogunlowo, A. S., Chancellor, W. J., and Gass, J. R.,** Dual-fueling a small diesel engine with producer gas from agricultural residues, Paper ASAE 79-1051, St. Joseph, Mich., 1979.
5. **Picken, D. J., Harris, L., and Collins, O. C.,** *Application of Biogas to Commercial Automotive Vehicles,* 2nd, E.C. Conference on Energy from Biomass, Strub, A., Chartier, P., and Schlesser, G., Eds., Applied Science, New York, 1982, 502.
6. **Ortiz-Cañavate, J., Hills, D. J., and Chancellor, W. J.,** Diesel engine modification to operate on biogas, Transactions of the ASAE, 1981, 808.
7. **Ahlers, R.,** *Electrical Control of a Cogeneration Set in Parallel Maine Operation or in Stand-By Operation,* 2nd. Conference on Energy from Biomass, Strub, A., Chartier, P., and Schlesser, G., Eds., Applied Science, New York, 1982, 517.
8. **Ashare, E.,** Analysis of systems for purification of fuel gas, in *Fuel Gas Production from Biomass,* Wise, D. L., Ed., CRC, Press, Boca Raton, Fla., 1981.
9. **Tennyson, R. N. and Schaaf, R. P.,** Guidelines can help choose proper process for gas-treating plants, *Oil Gas J.,* 1, 10, 1977.
10. **Anthony Henrich, R.,** Purification of digester gas to natural gas or vehicle fuel, presented to 3rd. Int. Symp. Anaerobic Digestion, Boston, August 14 to 19, 1983.
11. **Sáez, A. C., Reinke, G. S., and Alkalaj, D. L.,** Effect of fuel injection point on biogas dual fuel natural suction diesel performance, Proc. 3rd Int. Symp. Anaerobic Digestion, Boston, August 14 to 19, 1983, 379.
12. **Picken, D. J. and Hassaan, H.,** *Factors Affecting the Life of Engines Operating on Biogas,* 2nd, E.C. Conference on Energy from Biomass, Strub, A., Chartier, P., and Schlesser, G., Eds., Applied Science, New York, 1982, 506.
13. **Sáez, A. C., Reinke, G. S., Pischinger, F., and Schmillen, K.,** Joint Project on Application of Biogas as Fuel for Diesel Engines, University of Aachen, Germany, DFG-CONICYT, Unpublished data, 1985.
14. **Sáez, A. C., Caravajal, F. G., Reinke, G. S., Núñez, P. M., and Mesa, J. O.,** Studies on Generation and Use of Alternative Fuels, Research Report No, 842501 to UTFSM, 1984.
15. **Alkalaj, D. L., Sáez, A. C., Moltedo, C. M., Fuenzalida, D. O., and Henríquez, J. M.,** Project of a Pilot Plant of The Thermophilic Digester-Diesel Engine Total Energy System, Report No. 468/82 presented to Chilean National Fund for Scientific and Technological Development, March, 1984.
16. **Sáez, A. C., Reinke, G. S., Alkalaj, D. L., and Mesa, J. O.,** Biogas Application in Diesel Engines, Report to UTFSM and Municipality of Santiago, Chile, May, 1984.

Chapter 2

METHANE PRODUCTION BY CONTINUOUS DIGESTION OF FARM WASTES

J. P. Blanchard and T. A. Gill

TABLE OF CONTENTS

I.   Abstract..42

II.   Introduction...43
      A.   Methane Fermentation...43
      B.   Anaerobic Digestion Applied to Livestock Wastes 43
            1.   The Benefits of Anaerobic Digestion for On-Farm
                  Application 43
            2.   Energy Balance 44
            3.   Economics 45
                  a.   Capital Costs 45
                  b.   Maintenance Costs 45
            4.   Process Control Expertise...............................46

III. Assessment of Digester Performance.....................................47

IV.   Advances in Methane Reactor Process Design 48
      A.   Direction of Innovations.......................................48
      B.   Substrate Supply and Accessibility 49
            1.   Nutrient Balance 49
            2.   Substrate Accessibility.................................50
            3.   Separated Phase Reactors 52
      C.   Inhibition and Toxicity 53
            1.   Concentration, Antagonism, and Acclimation..............53
            2.   Cation Inhibition.......................................54
            3.   Heavy Metal Inhibition 56
            4.   Volatile Acids Inhibition...............................57
            5.   pH Inhibition 57
            6.   Ammonia Inhibition 58
            7. H_2 Inhibition..58
            8.   Sulfide and Sulfate Inhibition 59
            9.   Miscellaneous Toxic Materials...........................60

V.   Innovations in Reactor Designs 60
      A.   Biological Considerations for Biomass Retention................60
            1.   Increased Biomass Densities 60
            2.   Floc Formation 61
            3.   Attached-Film Formation 61
      B.   Retained Biomass Reactor Designs...............................62
            1.   Floc Based Designs 62
                  a.   Anaerobic Contact Process 62
                  b.   Sludge Blanket Designs 64
                  c.   Upflow Anaerobic Filters..........................65
            2.   Fixed-Film Reactor Designs 67

      a.   Stationary Fixed-Film Reactors.............................67
      b.   Expanded and Fluidized Beds (Surface Films) 69
      c.   The Use of Porous Biomass Support Particles.............72

VI.   Digester Kinetics..72
    A.   Basic Model for Microbial Growth and Substrate Removal 72
      1.   Microbial Growth and Substrate Assimilation....................72
      2.   Kinetics of a CSTR with a Simple Microbe and Substrate
            System..74
          a.   Cell Mass Balance...74
          b.   Substrate Mass Balance 74
      3.   Kinetics of the Anaerobic Digestion of Complex Wastes 76
      4.   Plug-Flow Kinetics 78
    B.   Retained Biomass Reactor Kinetics....................................79
      1.   Anaerobic Contact Process Kinetics............................79
      2.   Kinetics of Flocculent or Attached-Film Systems 81
    C.   Development of a Dynamic Model 83

VII.   Design of a Retained Biomass Reactor for Farm Wastes........................86

VIII.   Summary 91

IX.   Glossary of Terms 93

References..93

I. ABSTRACT

The benefits of on-farm application of the anaerobic digestion of livestock manure and crop wastes are outlined. The major obstacles preventing widespread adoption of this technology appear to be the high initial investment for construction coupled with a long payback period and the need for skilled operator expertise to prevent reactor failure due to the process variability inherent in a farming situation. Advances in two major areas (process intensification and development of a dynamic kinetic model) are overcoming these problems. A review of advances in understanding the mechanisms of substrate supply and accessibility and of the complex interrelated effects of various naturally occurring inhibitory agents in an anaerobic reactor are presented. A comprehensive understanding of these factors is required for the development of a dynamic model. A review of anaerobic digestion kinetics is also presented. The development of a dynamic model could lead to automated reactor control reducing the likelihood of process failure and relieving the farmer of the burden of providing skilled expertise. A review is also given of advanced reactor designs based on systems for retaining active bacteria. These retained biomass reactors are just beginning to be adapted to handle high particulate wastewaters such as livestock wastes and have greatly reduced the HRT required for anaerobic digestion. Shortened HRT's reduce the volumes required for reactor vessels which in turn reduces the initial cost for construction. Anaerobic digestion of agricultural wastes is destined for wider application with the development of high-rate retained-biomass reactors suitable for high particulate wastewaters and with the improvement of a dynamic model that will result in increased levels of automated reactor control.

II. INTRODUCTION

A. Methane Fermentation

Methane production from the fermentation of organic materials requires the concerted action of a symbiotic population of facultative and obligate anaerobic bacteria. Toerien and Hattingh[1] recommended that the symbiotic bacterial population could be conveniently divided into two major groups: "acid forming" bacteria and "methane forming" bacteria. The group that decomposed the initial organic substrate to produce a mixture of volatile fatty acids, hydrogen, ammonia, carbon dioxide, and less significant quantities of other metabolic by-products were classified as acid forming bacteria. Methane forming bacteria referred to the groups that converted substrates such as acetic acid, hydrogen, and carbon dioxide to methane.

More recent studies have labeled three major groups of symbiotic bacteria.[2] The first groups, called fermentative or acidogenic bacteria, degrade carbohydrates (largely cellulose) lipids, and proteins to form alcohols, organic acids (lactate, succinate, etc.), and long chain fatty acids. The second group, called acetogenic bacteria, produce acetate, carbon dioxide, and hydrogen from the end products of the first stage. In the last stage, methanogenic bacteria produce methane and carbon dioxide via mainly the following two reactions.[3] Acetic acid fermentation:

$$CH_3COOH \rightarrow CH_4 + CO_2$$

and, CO_2 reduction:

$$CO_2 + 4H_2 \rightarrow CH_4 + 2H_2O$$

The acidogenic and acetogenic bacterial groups are the most prolific, while the methanogens have the slowest growth rate.[4] A healthy population of methanogenic bacteria prevents the accumulation of volatile fatty acids that could lead to digester failure.[5] A balanced state of dynamic equilibrium among the anaerobic groups can be most easily maintained within a controlled environment. A wide variety of physical and biochemical parameters influence the types and numbers of bacteria that constitute a stable methane fermentation. Environmental control for the continuous generation of methane gas requires that parameters such as the concentration of microbial nutrients, pH, substrate concentration, the numbers of viable microorganisms, and the total volatile acid content be maintained within acceptable limits.[4]

B. Anaerobic Digestion Applied to Livestock Wastes

1. The Benefits of Anaerobic Digestion for On-Farm Application

Anaerobic digestion of farm wastes has a number of advantages not offered by other methods of waste management. The effluent from anaerobic digestion has a greatly reduced solids content, less offensive odors, and fewer handling problems.[6] Fermentation results in an effluent with lower solids content than aerobic digestion since more of the organic material is converted to gas and less to cell mass.[7] Approximately 5% of the degraded organic waste material is converted to biological solids during anaerobic digestion, while most of the remainder is converted to biogas.[8] In comparison, the biomass yield from aerobic digestion is in the range of 50% of the mass of converted organic substrate. It has been frequently demonstrated that biological oxygen demand (BOD) reductions greater than 80% are achievable from anaerobic digestion of livestock manures.[9]

The value of anaerobic digester effluent compares favorably with raw manure as a crop fertilizer because the inorganic nutrients (minerals) pass through the digester undepleted. The amount of nitrogen available for plant assimilation is increased by the conversion of protein nitrogen to ammonia.[10] However, if the effluent is stored before use, the ammonia may be subject to volatilization to the atmosphere. Pathogenic bacteria such as *Salmonella* sp. and *Brucella* sp. are destroyed during anaerobic fermentation, and livestock can be grazed on pastures that have been spread with digester sludge sooner than would be acceptable if raw manure was used.[9] A major advantage of anaerobic digestion is the production of biogas that is typically about 60% methane with a fuel value of 2.24×10^7 J/m^3.[11] For example, a 450 kg dairy cow produces approximately 3.9 kg of volatile solids as waste per day.[12] From this, a methane yield in the range of 0.35 to 0.81 m^3, equivalent to the fuel value of 0.2 to 0.5 ℓ of diesel oil, could be expected.[13]

2. Energy Balance

The major energy inputs for continuous mesophilic anaerobic digesters are heating and mixing.[9] Various species of anaerobic bacteria thrive at different temperatures. Three groups of anaerobes are classified according to their optimal temperature ranges for growth.[11]

1. Psychrophilic bacteria: 5 to 35°C
2. Mesophilic bacteria: 18 to 45°C
3. Thermophilic bacteria: 45 to 85°C

It is generally accepted for temperate climates that operation in the mesophilic range is the most efficient compromise between the required thermal inputs and losses and the rate of reaction.[14] The choice of a temperature to optimize the net energy balance of a digester in the mesophilic range depends upon the digester design and operational parameters. Hawkes and Horton[15] recommended the use of ''net energy production'' as a criterion for selecting the operating temperature. They found that increases in gas production with digester temperature were not linear. For example, the extra methane produced by raising a digester's temperature a few degrees (say from 30 to 35°C) might be offset by increased heat demands. A separate assessment of net energy production vs. temperature for each particular case was recommended. The slower bacterial growth rates at lower temperatures may be partially overcome by increasing the biological solids retention time (BSRT) with a recycled or retained biomass design.[16]

In temperate climates, efficient operation of digesters maintained at 30 to 35°C during cold seasons is a challenge. Proper insulation in the construction helps to minimize heat losses from the reactor.[17] A major thermal input that is more difficult to mitigate is the heating load required to raise the feed slurry from ambient to the reactor temperature.[18] The amount of thermal input to the feed slurry is partially dependent on the conservation of heat in the manure collection system. The addition of wash water also might increase the heating load. The feed slurry may require heating by as much as 30°C in winter months.[13] Heat reclaimed from the digester effluent would be a desirable form of energy conservation, but is complicated by the difficulties associated with handling and pumping manure, by the tendency for heavy surface fouling of heat exchanges by manure, and by the small temperature differences between the loading and effluent that seldom exceed 30°C.

The level of adequate mixing energy is variable depending on a large variety of design factors.[19] A continuous anaerobic digestion system requires sufficient mixing to provide adequate bacterial access to the substrate, to maintain a uniform digester temperature, to ensure that the loading displaces well mixed liquor as effluent, and in some cases to prevent crust formation.[20] The plug-flow digester design is an exception since it requires little or no mixing. Either mechanical agitation, head space gas recycling, or liquid pumping can be used for digester mixing.[21] Zoltak and Gram[22] found the level of mixing in a digester correlated well with the actual amount of power input regardless of the mixing method. A report by the U.S. Environmental Protection Agency[23] cited power requirements of 5 to 8 W/m^3 of digester volume for strong mixing by gas recirculation.

Positive energy balances have been reported during perpetual operation of present day methane generation systems utilizing agricultural wastes.[24] In a 4.5 year Canadian study at the University of Manitoba,[25] four 3000 ℓ continuous stirred-tank reactor (CSTR) anaerobic hog manure digesters were continuously operated at different temperatures, hydraulic retention times (HRT) and loading rates. All four digesters had positive energy balances even during the winter months. The potential exists for further improvements to energy balances using reactor designs with increased process intensities. A number of recent on-farm anaerobic reactor installations designed for high rate operation have demonstrated surplus energy productions during winter conditions.[26-29]

3. Economics
a. Capital Costs

The cost for digester construction depends in part on the design and more importantly, the size. A plug flow design developed at Cornell University claimed a relatively inexpensive construction cost.[7] The installation costs for plug flow digesters of 150 to 1000 m^3 capacity were calculated from the data of Jewell et al.[7]

$$IC = 725 \ V^{0.52} \tag{1}$$

where: IC = installation cost ($U.S.); and V = digester volume (m^3).

Construction of a mechanically mixed CSTR with a long HRT would be a relatively more expensive design. The accepted range of HRT for CSTR's for animal manure fermentation is 7 to 30 days.[6] Bacterial wash out is a risk at an HRT of less than 10 days, while the slight increases in gas production and BOD reduction obtained using an HRT greater than 30 days do not justify the added expense for the construction of a larger reactor. A survey of construction costs by Hashimoto et al.[24] for a number of on-farm continuous stirred-tank reactor's ranging from 10^3 to $> 10^4 m^3$ gave the following minimum and maximum cost curves ($U.S.) as a function of volume (m^3):

$$IC_{min} = 2970 \ V^{0.7} \tag{2}$$

$$IC_{max} = 5320 \ V^{0.7} \tag{3}$$

The development of retained biomass anaerobic digesters for agricultural wastes has the potential to reduce HRT from the 10 to 20 days required for conventional anaerobic digesters to less than 1 day.[30] Installation costs could obviously be reduced by a factor greater than 10 while retaining the benefits of stable operation and high levels of BOD reduction that result from long BSRT's.

b. Maintenance Costs

The annual operating costs for an on-farm anaerobic reactor typically include costs for electricity, water, maintenance, taxes, insurance, repairs, and parts replacement. Hashimoto et al.[24] estimated annual costs to be less than 10% of the initial construction cost. The economic feasibility of farm scale anaerobic digesters has been a matter for debate. Differences arise due to variable factors such as the local value of fuel or electrical energy, reactor design (construction cost), the pay back term used for economic analysis, and the estimates of labor costs for digester operation. Hashimoto et al.[24] concluded from an economic survey that farm waste fermentation was economically attractive for a sufficiently large-scale installation. Persson and Bartlett[17] concluded that farm digesters could yield a positive financial return provided the energy output was fully utilized for heating needs on the farm. They did not specify a size restriction on the installation. Stafford and Etheridge[26] found that farm scale digesters of the anaerobic contact design (Section V.B.1.a) were economically feasible in the U.K., at the time of their study, but had a return on investment of less than 10%. However, they stipulated that if a 10% per annum increase in electricity costs over a 6 year period occurred, then the return on investment would become quite attractive. All three of these studies included labor cost allowances. The minimum installation size restriction recommended by Hashimoto et al.[24] was in large part due to allowance for the wages and benefits of at least one full time operator for the reactor. Persson and Bartlett[17] assumed that operation could be provided by existing farm labor. Stafford and Etheridge[26] made allowances for cost of additional labor to operate the digesters.

It would appear that the economic value of methane production can pay back capital and maintenance costs for on-farm anaerobic digesters, but that labor costs are more difficult to determine and make the overall economic viability of farm installations questionable and dependent on the size of the operation especially if full time skilled labor is required. Labor costs depend partly on the level of digester control and monitoring required. The level of manual monitoring and control required are in turn a function of the process stability of the reactor, the mechanical dependability of the materials handling and mixing equipment, and the level of automation incorporated into monitoring and process control.

As noted previously, development of retained biomass reactor designs for farm wastes could substantially reduce the digester volumes required for a given application resulting in reduced initial construction costs. Process stability would also be increased due to the increases in BSRT.[31]

4. Process Control Expertise

Recent studies indicate that with new technical developments, on-farm livestock waste methane fermentation is potentially economically attractive. However, the requirement for operator time and expertise for process monitoring and control, and the lack of a well defined dynamic process control model are major obstacles to the present widespread application of anaerobic digesters for farm waste management and renewable energy production.[32] Monitoring digester performance entails a number of chemical and physical measurements that require technical expertise and in some cases access to laboratory facilities. Quite understandably, farmers who are busy may be reluctant to install methane digesters because of the time required for initial process optimization and the on-going adjustments for process deviations. The unpredictability of performance and the required level of expertise are major deterrents for most farmers even though a well designed fermentor operating at its stable design capacity can be an economically attractive form of livestock waste management. The very real risk of digester failure[24] is another deterrent. Failure may often occur as a result of "shock" loading or due to the introduction and accumulation of inhibitory substances in the reactor that upset the delicate balance between nutrient assimilation and bacterial reproduction.

Buvet[33] concluded from an economic analysis of anaerobic farm waste digestion that the labor costs of a full time analyst and one assitant for monitoring and process control could only be justified for a minimum digestion capacity of 10^4 m^3. This restriction obviously precludes the economic viability of most single farm units and it implies that perhaps some form of cooperative effort on a regional basis would be necessary to economically provide monitoring and control expertise. The level of operator expertise for process control and the risk of digester failure could be reduced by the development of a comprehensive mathematical model to predict the dynamic behavior of an anaerobic digester during periods of process deviation and during digester start-up. Work on the development of a dynamic model that utilizes as input a few basic parameters regularly measured in anaerobic reactors has been reported.[32,34-37] These models incorporate the interrelated effects of process parameters such as loading rate, temperature deviation, growth and decay rate of the microorganisms, by-product inhibiton (i.e., volatile acids, ammonia, etc.), toxin accumulations, and pH fluctuations. The analytical solution of the Monod type of microbial growth and substrate utilization equations[38-39] is only possible for the steady-state case in which bacterial numbers and substrate concentrations are constant. The steady-state equations are useful for selection of a digester design but do not apply to situations in which bacterial growth rates and substrate utilization are not constant. Mathematical techniques such as numerical analysis may be used to model the unsteady-state. These have only become feasible with the development of high speed computers. The dynamic models are not fully developed to account for all the process variables in a farm waste anaerobic digester, but these models have generated predictions of dynamic behavior based on some of the more common changes in process conditions of actual digesters.[32] Information necessary for the development of accurate models includes the physical and biochemical characteristics of livestock wastes, substrate conversion coefficients and microbial inhibition factors. These data will be outlined in the following section.

III. ASSESSMENT OF DIGESTER PERFORMANCE

The lack of consistency in reporting the results of laboratory and field studies of anaerobic reactors makes it difficult to compare past studies in a comprehensive manner. Demuynck et al.[40] have suggested that the reporting of standardized parameters and measurements would be beneficial for correlating results from different anaerobic digestion projects. These included:

1.   A standardized set of operational and performance parameters.
2.   Data on the physical and biochemical characteristics of the digester feedstock.
3.   A detailed description of the digester design and construction.
4.   An overall energy balance for the digester with an economic analysis.

Eleven operational and performance parameters for evaluating anaerobic digestion studies were recommended.[40]

1.   Substrate concentration in the feedstock
2.   Substrate concentration in the effluent
3.   The rate of biogas production
4.   The concentration of methane in the headspace gas
5.   Digester pH
6.   Operating temperature
7.   HRT
8.   Loading rate
9.   Methane yield
10.   Efficiency
11.   Conversion

The above authors defined ''yield'' as the methane produced (in liters) per gram of substrate. The ''efficiency'' was defined as the amount of methane produced per unit substrate utilized by the digestion and the "conversion", as the amount of substrate removed per unit of substrate introduced.

These eleven basic parameters describe an anaerobic reactor's performance. Additional information about feedstock, digester contents, and effluent are required to rationalize and explain the performance. Careful characterization of the feedstock used in anaerobic fermentation studies would facilitate the accumulation of correlatable data that could contribute to the refinement of an accurate and comprehensive predictive model for methane fermentations. Factors related to the concentration and availability of substrate for bacterial growth, and to the concentration of potentially inhibitory substances or precursors of these substances all have an influence on the performance level to be expected from a particular digester. Therefore, these factors must be considered directly or indirectly in any comprehensive predictive model of biomethanation.

Parameters affecting the digestion feedstock are: the concentration of organic substrate [usually measured in terms of volatile solids (VS), oxygen demand (COD and BOD)]; the actual biodegradability of the feedstock; the nutrient composition including levels and ratios of the macronutrients (C,N,P,S), the micronutrients such as the alkali cations (Na, K, Ca, Mg), the heavy metals (Zn, Cu, Ni, Co, Fe); the the proportions of the major biological compounds that comprise the volatile solids (i.e., carbohydrates, proteins, and lipids). Callander and Barford[41] recommended that any new waste material considered for anaerobic digestion should be thoroughly tested for all known nutrients.

The relevant biochemical parameters of the digester's contents and effluent for consideration in a dynamic model include such indicators as: the level of volatile solids or COD remaining in the effluent; digester pH; volatile acids concentration; alkalinity; rate of gas production and its composition; active biomass density in the digester and its turnover rate; concentration of such substances as ammonia, alkali cations, heavy metals, sulfate, sulfides, and H_2 and any miscellaneous toxins such as detergents, disinfectants, and antibiotics. A more detailed review of the feedstock and digester liquor parameters listed above will be given in Section IV.

A number of different anaerobic reactor designs have been proposed to enhance biogas production. The general categories of anaerobic reactors in use or under study are: batch reactors, plug flow reactors, CSTR, CSTR with biomass recycle, and continuous reactors employing bacterial retention systems. Specifying the reactor design in the report of digester performance is necessary to the development of a kinetic model.[38] As an example, determination of a kinetic model for an anaerobic digester design requires a knowledge of the BSRT.[42] In a traditional CSTR, the BSRT is equal to the HRT, but in a biomass recycle or retention system this is not so.[43] The determination of BSRT for biomass recycle systems can be obtained with a materials balance for microbial mass on the loading influent, the effluent stream, the recycle stream, and the wastage flow from the recycle stream.[44] For a retained biomass system, the determination of BSRT is more difficult. As well, the kinetic model of reaction rates in retained biomass reactors may be further complicated by nutrient diffusional limitations in bacterial films or flocs.[45]

IV. ADVANCES IN METHANE REACTOR PROCESS DESIGN

A. Direction of Innovations

Recent innovations in anaerobic digestion technology have resulted in improved digester performance and stability. These advances have been gained using two approaches:[41] increased biomass activity and increased biomass density per unit reactor volume. The first of these approaches to improve digester performance will be discussed in this section. The design of anaerobic reactors with improved reaction rates due to increased biological activity has resulted from advances in the biochemistry and microbiology of methane fermentation.

Progress has been gained from the elucidation of the nutrient requirements for a balanced anaerobic digestion, and the interrelated inhibitory behavior of certain metabolic byproducts and toxins that can accumulate during substrate conversion. This knowledge has led to improved techniques to maintain optimally balanced microbial growth conditions. Section V will review the advances in anaerobic digestion technology resulting from innovative reactor designs that provide increased biomass density per unit reactor volume.

B. Substrate Supply and Accessibility

1. Nutrient Balance

Substrate composition plays a major role in the determination of the microbial ecosystem that comprises the complete fermentation. The organic and inorganic components as well as the physical characteristics of the substrate lead to a natural selection of those bacteria which best metabolize these components in a balanced system.[46]

Bacteria forming an anaerobic digestion require a sufficient and balanced supply of the major organic nutrients: carbon, nitrogen, phosphorous, and sulfur which are the building blocks of biological molecules. Elements not present in the proper ratios will become growth-limiting and inhibit efficient digestion. The required levels of phosphorous and sulfur are much less than the required levels of carbon and nitrogen. In the case of many wastes such as municipal sewage, livestock manure, or farm crop residues, the level of the major macronutrients are relatively well-balanced.[24] It is mainly specialized industrial or food processing wastes that may grossly lack one or more of these elements.

The nitrogen requirement is about 11% of the dry weight of cell mass produced, based on the formula, $C_5H_9O_3N$, for the average chemical ratio of biological cells.[8] The phosphorous requirement is approximately 2% of the dry weight of the cell mass produced. The demand for nitrogen in a stable anaerobic digestion is greater than the amount that would be predicted from cell mass balances. Excess ammonia in the digester contributes to the formation of ammonium bicarbonate which provides an effective buffer that stabilizes pH.[47] Most of the ammonia in an anaerobic digester is produced by the decomposition of protein. Livestock wastes are relatively high in organic nitrogen and produce correspondingly high levels of ammonia during fermentation.[24] As noted earlier, the cell mass produced from anaerobic digestion is typically about 5% of the volatile solids that are degraded; the rest of the volatile solids contribute mainly to gas production.[44] Using the average chemical ratios of nitrogen and phosphorous in cell mass cited above, the requirement for nitrogen and phosphorous for cell growth during anaerobic digestion of dairy manure would be approximately 0.026 and 0.005% of the weight of undiluted raw manure, respectively. These figures were calculated using the assumption that the biodegradability of dairy manure is approximately 40% of the total solids,[48] and that the level of total solids in fresh dairy manure is approximately 12%.[49] Table 1 presents the amounts of nitrogen, phosphorous, and potassium typically found in various livestock manures. It can be seen that the levels of nitrogen and phosphorous are approximately 10 times greater than those required according to the values calculated from cell mass production during anaerobic digestion.

Several researchers[51-52] found that the typical carbon to nitrogen ratios of cow dung were lower than carbon to nitrogen ratios that gave the maximum gas production in anaerobic digestion trials, and that mixtures of crop waste and cow dung adjusted to give carbon to nitrogen ratios of approximately 30:1 yielded the maximum gas production rates. Ghose and Das[51] used mixtures of cow dung with various combinations of rice husks, and finely ground water hyacinth and algae. Hills[52] used cow dung mixed with chopped barley straw. In Hills[52] study, the carbon contents were evaluated as ''nonlignin'' carbon, based on measurements of the total organic carbon minus the lignin carbon.

The biological growth yields from the fermentation of different wastes vary depending on the ratios of carbohydrates, proteins, and lipids that comprise the degradable fraction of

Table 1

**NITROGEN/PHOSPHOROUS/POTASSIUM RATIOS FOR
VARIOUS FARM MANURES (FROM U.S. AND U.K. SOURCES)**

	% of raw weight				
Animal	**Total solids**	**Total N**	**Total P**	**Total K**	**N:P:K (ratio)**
Swine feeder[a]	9.2	0.69	0.23	0.45	1.53:0.51:1.
Swine feeder[b]	9.5	0.63	0.16	0.19	3.32:0.84:1.
Poultry layer[a]	25.2	1.36	0.53	0.58	2.34:0.19:1.
Poultry layer[b]	23.0	1.50	0.55	0.41	3.36:1.34:1.
Beef feeder[a]	11.6	0.57	0.19	0.42	1.36:0.45:1.
Beef feeder[b]	11.1	0.51	0.16	0.47	1.09:0.13:1.
Dairy cow[a]	12.7	0.49	0.09	0.33	1.48:0.27:1.
Dairy cow[b]	11.1	0.51	0.06	0.47	1.09:0.13:1.

[a]   data from ASAE[49]

[b]   data from O'Callaghan et al.[50] for animals in U.K.

the volatile solids.[31] The production of anaerobic bacterial cell mass per unit mass of ultimate biological oxygen demand (BOD_L) was found to be greatest for carbohydrates, intermediate for proteins, and least for lipid substrates. Cowley and Wase[9] cited the theoretical volumes of methane that could be produced from the complete fermentation of the three major carbon sources: carbohydrates, proteins and lipids, as 0.37, 0.49, and 1.04 m^3 (STP)/kg of initial dry substrate, respectively.

Other nutritional factors are required for optimum cell growth in low concentrations or trace quantities.[46] Included among these are alkali salt cations such as magnesium, calcium, potassium, and sodium; as well as metals such as zinc, iron, cobalt, copper, molydenum, and manganese. Some of these act as cofactors for microbial enzyme reactions.

2. Substrate Accessibility

The fraction of biodegradable material in the feedstock is an important factor for the accurate application of a substrate removal and microbial growth model. The level of biodegradability is quite variable depending on the type of organic waste. Morris et al.[48] found that only 42.5% of the volatile solids of dairy manure used in kinetic studies was biodegradable.

It is generally accepted that the rate-limiting step for the anaerobic digestion of many different wastewaters is the conversion of acetic and propionic acids to methane.[43] McCarty[31] found that the breakdown of long chain fatty acids had a reaction rate similar to that for methanogenesis of acetate and propionate. However, this is of little concern when dealing with animal manures which typically have a lipid content of less than 10% of the total solids.[53] Consideration of the composition of plant material that has passed through the digestive tract of livestock is relevant to an assessment of the biodegradability and rate of conversion that might be expected during anaerobic digestion. Ruminants are usually fed diets containing relatively high levels of roughage (cellulose fiber) which is hydrolyzed by the rumen flora of the gut. Pigs and chickens are fed less roughage, but digest little cellulose, producing manures with significant levels of cellulosic fibers.[11] Differences in the anaerobic digestibility of various livestock manures are reflected in the data of Table 2.

The relatively low levels of volatile solids converted in dairy manure fermentation would be expected due to the higher ratio of biologically inert woody plant tissue in the organic solids. Cellulose, a major source of carbon in plant material, is totally degradable to acetate and thus, to methane and carbon dioxide; lignin, a structural component of plant tissue, is completely inert to anaerobic bacterial attack.[54] The digestibility of cellulose in plant fiber has been inversely correlated to the lignin content.[55] Much of the cellulose in manure is

Table 2
**COMPARISON OF VS REDUCTION FOR LIVESTOCK WASTES
IN MESOPHILIC ANAEROBIC DIGESTERS.**

Animal	Number of examples	Detention time (days)	% VS reduction	CH$_4$ production (m^3/kg VS removed)
Dairy[a,b]	17	10—30	32.0 ± 11.8	0.38 ± 0.17
Swine[a]	6	10—22	52.7 ± 6.0	0.51 ± 0.15
Poultry[a,c]	3	10—30	67.5 ± 0.2	0.40 ± 0.17

[a] Data from a collection of digestion performance characteristics cited by Midwest Plan Service.[11]
[b] Data from a collection of digestion performance characteristics cited by Chynoweth et al.[59]
[c] Data from Klein.[60]

held in a lignin matrix resulting in increased resistance to enzymatic attack.[52] Anaerobic biodegradability has been improved by pretreatment with heat, strong acid, strong base, and particle size reduction.[56-58] However, the economic viability of these pretreatments for improved anaerobic fermentation is questionable.

Equipment to separate manure slurries into solid and liquid fractions has recently been introduced for agricultural waste management. Separation techniques were originally devised for more efficient handling, storage, and treatment of manure. It has also been noted that the process of manure separation has advantages applicable to anaerobic digestion. Liquid/solid separation has effects similar to those of particle size reduction. During the separation process, the large particles are removed from the system (often by screening) before digestion. Most separation systems were designed to remove the fraction of largest fibrous solids that typically comprise 40 to 50% of the total solids in the original manure slurry. Approximately 60% of the total solids and 70% of the COD in whole manure have been found to pass through mesh sizes ranging from 0.044 to 0.25 mm.[11] Harper et al.[61] found that the liquid fraction of swine and cattle manure passing through a 60-mesh (0.25 mm) screen had a rate of biodegradation almost twice that of whole manure. They found a semi-log relationship between the COD (mg/ℓ) and digestion time for a batch anaerobic digestion. This implied first order kinetics and the following equation for COD removal efficiency (E) was found to be applicable:

$$E = KT/(1 + KT) \tag{4}$$

where: E = COD removal efficiency; K = COD removal rate constant (day^{-1}); and T = detention time (days). The detention times required for 50% COD removal of separated dairy or swine manures by anaerobic digestion were approximately one half the detention times required for 50% COD removal from whole manure. It was notable that the fraction of large solid particles separated from the whole manure was stable for extended storage periods and produced little odor.

The potential influence of the livestock feeding ration on the digestion characteristics of manure can be inferred from a study by Frecks and Gilbertson.[62] Two groups of beef cattle were fed different rations; one, a high roughage diet, and the other, a highly concentrated ration. They found that the high roughage diet yielded 22% more fecal particles retained by small sieve sizes (0.0197 to 0.105 mm), although the total solids content of the feces and urine from both manure samples were the same. As well, the manures produced from the highly concentrated diet had higher volatile solids contents than did the high roughage ration manures. These findings suggest that a more highly concentrated diet may produce a manure with a faster rate of anaerobic digestion and a greater yield of methane because of the smaller average particle sizes and higher volatile solids contents.

Water content of farm wastes is generally dictated by practical considerations of materials handling. Efficient digester mixing is an important consideration if the waste is to be used for methane production. For these reasons, Loehr[63] recommended the solids content of anaerobic reactors be less than 10%. However, enzyme reaction kinetics would predict highest production rates at highest practical solids concentrations.[64] Researchers at Cornell University developed a ''dry fermentation'' system for crop residues and animal wastes that operates with feedstocks with up to 30% solids.[65] This design has been successfully operated in the thermophilic range in a batch reactor initially seeded with anaerobic municipal sewage sludge. In some instances, when toxic substances have accumulated to inhibitory levels in anaerobic batch reactors, digester failure was prevented by dilution with water to lower the concentration of the inhibitory substance.[31]

3. Separated Phase Reactors

The biomethanation process is accomplished by a symbiotic population of various groups of bacteria. Soluble and particulate carbohydrates, proteins, and lipids are sequentially degraded by a series of different types of bacteria. The organic solids are initially hydrolyzed and converted to intermediate products such as volatile acids, carbon dioxide, hydrogen, ammonia, and sulfides.[2] Acetic acid typically accounts for approximately 70% of the short chain volatile acids formed.[66] Hence, the term acetogenesis was applied to the stage of methane fermentation producing acetic acid. The intermediate volatile acids are then converted by a separate group of bacteria called methanogens into mainly methane and carbon dioxide. The substrates and environmental conditions required to achieve optimal conversion rates are different for the acetogenic and methanogenic groups. The kinetics of their rate limiting microbial reaction steps are also different.[67]

Various physical, biochemical, and kinetic approaches have been used to separate the complete anaerobic digestion into acid forming and methane forming stages. Borchardt[68] employed a dialysis technique to separate these phases (acetogenic from methanogenic). This technique was suitable mainly for lab-scale studies of isolated methanogenesis. Other researchers have separated anaerobic digestion into two phases by methods such as the addition of inhibitors or the adjustment of pH in the first stage to promote or discourage growth of the methanogens.[69,70] Successful phase separation has also been achieved by utilization of the differences in biochemical reaction kinetics between the acid forming stage and the methane forming stage.[69,71-73] Massey and Pohland[67] achieved phase separation by the use of an appropriately short HRT for the first stage. This prevented establishment of the methane forming bacteria due to their slower growth rate. The growth of acid forming bacteria in the second stage was suppressed since little of the substrate required for their growth remained in the effluent from the first stages. This type of phase separation allowed the dual optimization of environmental conditions for the acid formers and methane formers, and permitted the application of kinetically selected HRT's for each phase. Massey and Pohland[67] determined a series of kinetic constants for both the acid forming phase and the methane forming phase. The feedstock to the first stage was a soluble substrate. They concluded that a HRT of about one day was appropriate for the acid phase and 15 to 20 days was suitable for the methane phase. They also found that the settling and flocculating properties of acid forming bacteria allowed biomass recycle for the first phase, but settling of the methanogens was too slow to make efficient use of biomass recycle for the methane phase.

Ghose and Bhadra[71] found that the utilization of separated phase anaerobic digestion for cow manure improved the overall performance and efficiency of volatile solids reduction and energy production. They employed a series of four acid phase digesters each with a

HRT of 18 hr followed by a methane digester with a HRT of 288 hr. Volatile solids reduction and methane yields were approximately 1.5 times greater than a comparable single stage batch digestion, and overall energy recovery was doubled. It was concluded that the increase of volatile acid levels in the methane phase under optimal operating conditions led to increased numbers of methane formers, resulting in increased levels of ATP and increased generation of hydrogen carrier in that phase. This led to increased conversion of CO_2 to CH_4. Also, Massey and Pohland[67] found that recycling a common gas phase through both stages of separated phase digestion appeared to increase the methane yield and volatile solids reduction probably because the H_2 produced by the acid phase was available for CO_2 reduction to methane in the methanogenic phase. Other advantages could be expected from a separated phase design. The use of a separated phase digester would allow the individual optimization of parameters such as pH, substrate concentration, methods of inhibition control, and operating temperature for each phase.

Phase separation of anaerobic digestion has enabled the utilization of several retained biomass reactor designs for the digestion of highly particulate wastewaters such as livestock manure. Without phase separation these designs were normally restricted to treatment of solubilized wastewaters. Pig manure was efficiently digested in a separated phase anaerobic treatment system utilizing an upflow anaerobic filter (Section V.B.1.c) with a 3 day HRT for the methanogenic phase.[30] The pig manure slurry was subjected to hydrolysis over a 12 to 15 day period in storage tanks at ambient temperatures. Gravity settling of the less biodegradable solids occurred during this time and the supernate which contained less than 1% suspended solids with a COD between 10,000 and 60,000 mg/ℓ, was fed to an upflow anaerobic filter. This system resulted in COD reduction and methane yields that were often greater than those reported in the literature for conventional reactors using 10 to 25 day HRT's. The anaerobic filters provided improved stability against sudden loading rate changes and could be quickly restarted after short shut-down periods without loss of efficiency. Smith et al.[74] also demonstrated the utilization of an initial acidogenic stage followed by an anaerobic upflow filter treatment as the second stage for the digestion of poultry and swine manure slurries. This design also had the facility to maintain stability during periodic loading. Norrman and Frostell[75] designed a two stage anaerobic digestion system employing an anaerobic contact reactor (Section V.B.1.a) as the first stage hydrolysis tank, and an upflow anaerobic filter as the second stage methane reactor. The pH of the hydrolysis reactor was approximately 4.7, while the pH of the second stage anaerobic filter rose to between 6.4 and 7.5. This system also demonstrated improved stability of operation compared to the traditional digester designs, and the lab scale studies indicated this system was probably capable of handling most kinds of biologically degradable wastewater.

C. Inhibition and Toxicity

1. Concentration, Antagonism, and Acclimation

Investigation of the role of inhibitory and toxic materials has led to better process control strategies for anaerobic reactors and techniques to maintain optimum microbial growth conditions. Common inhibitory conditions include: cation or salt inhibition, heavy metal toxicity, volatile acids inhibition, ammonia inhibition, sulfide inhibition, and in the case of livestock wastes, the introduction of cleaning agents and antibiotics to the reactor.

Many of the more common potentially toxic agents such as the cations Na, K, Ca, Mg, and the heavy metals are essential nutrients at low concentrations and become inhibitory as the concentration is increased during fermentation.[76] Past studies of anaerobic digestion have at times reported widely varying levels of inhibitory substances. This perhaps occurred because of the number of complex factors that influenced the severity of inhibition resulting from a given concentration of a toxic agent. Kugelman and Chin[77] outlined four major factors that influenced the degree of toxicity caused by a specific concentration of a particular toxic substance. These factors were antagonism, synergism, complex formation, and acclimation.

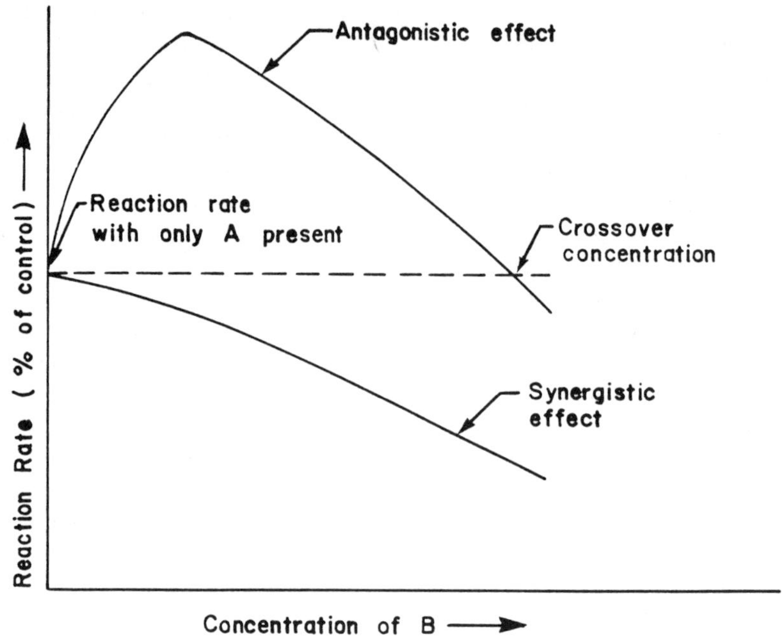

FIGURE 1.   Typical effects of antagonism and synergism in a two component system with the concentration of substance A held at a toxic concentration. (From Kugelman, I. J. and McCarty, P. L., Cation toxicity and stimulation in anaerobic waste treatment, *J. Water Pollut. Control Fed.*, 37(1), 97, 1965. With permission.)

Antagonism was defined as the reduction of the toxic effects from one substance by the presence of a second substance. Synergism was defined as an increase in toxicity from one substance by the presence of any concentration of another. Figure 1 illustrates typical antagonistic and synergistic behavior. Antagonism is shown as a decrease of inhibition from substance A by the presence of low levels of the antagonist B. As the concentration of the antagonist B is gradually increased, the level of inhibition increases until it exceeds the original level of inhibition with only A present. On the other hand, if the second substance B is synergistic, the rate of reaction begins to decline immediately as substance B is introduced to the system and declines further as the concentration of B is increased. Antagonistic and synergistic relationships are difficult to delineate in wastewater digesters since the concentrations of substances that exert antagonism or synergism are usually very low. The antagonistic and synergistic effects of combinations of light metal cations on methanogenesis has been studied by Kugelman and McCarty.[78]

Complex formation refers to the binding of soluble ions to form a less soluble salt that precipitates from solution. This is an effective way of lowering the concentrations of some solubilized toxic substances in an anaerobic reactor. The most common application of this phenomenon is the removal of heavy metals such as Cu, Ni, and Zn from solution by complex type reactions with sulfides.

Acclimation is the demonstrated ability of balanced anaerobic fermentations to accomodate themselves to gradually increased levels of inhibitory substances. Acclimation generally is the result of a gradual rearrangement of the metabolic processes of an organism to bypass the metabolic block caused by the toxic agent.[77]

2. Cation Inhibition

The toxic effect of salts on bacterial growth has been studied extensively early in this century.[79] The toxic effect of salts was largely attributed to the cations. Early studies on the inhibitory properties of cations such as Na^+, K^+, Ca^{+2}, Mg^{+2}, and NH_4^+ on anaerobic fermentation were generally conducted on sewage sludge digesters[80,81] where it was difficult to control and define the ionic composition of the reactor contents. This situation led to conflicting reports of the inhibitory levels of cations since the effects of antagonism and synergism were difficult to determine in such complex digestion media. Kugelman and McCarty[78] did studies of cation toxicity using synthetically derived substrate media which allowed observation of the effects of antagonism and synergism with controlled combinations of the light metal cations: Na^+, K^+, Ca^{+2}, Mg^{+2}, and NH_4^+. Ionic concentrations for maximum digestion efficiency were 0.01 *M* for the monovalent ions (Na^+, K^+, and NH_4^+), and 0.005 *M* for the divalent ions (Ca^{+2}, Mg^{+2}). The inhibition caused by excessive concentration of any one of these ions could be antagonized by the addition of the optimum level of at least one of the other cations of the group. Maximum antagonistic effects were achieved by the addition of several other cations at their optimum concentration. In some instances, the addition of certain other cations acted as synergists rather than antagonists. Kugelman and McCarty[78] suggested that if certain cations were enzyme activators, they would increase reaction rate until the enzyme was fully activated. Further increases in the cation concentration then would possibly result in the inhibition of other enzymes, thus decreasing the overall digestion rate. If a different cation were then added at its optimum level, it might reactivate or speed-up the blocked enzymes in an antagonistic manner. Kugelman and McCarty[78] found that if cation concentrations were increased gradually rather than in a single step, higher concentrations of cations could be tolerated because of acclimation. They obtained these results in simple well-defined media so that the specific concentrations they cited for specified level of inhibition applied for only one organic loading rate and BSRT. Kugelman and Chin[77] performed a cation inhibition study similar to that of Kugelmand and McCarty[78] to obtain information for the behavior of selected cation pairs on the rate of the acetate conversion reaction in anaerobic digesters. Their approach was to determine the values of five kinetic coefficients including a microorganism decay rate coefficient at various cation concentrations. These coefficients were determined using the following equations for substrate and cell mass concentrations.

$$S_1 = \frac{K_s(1 + k_d\,\theta_c)}{\theta_c(\hat{\mu} - k_d) - 1} \tag{5}$$

$$X_1 = \frac{(S_0 - S_1)\,Y}{(1 + k_d\,\theta_c)} \tag{6}$$

Where: S_1 = effluent substrate concentration (mg/ℓ); S_0 = influent substrate concentration (mg/ℓ); θ_c = biological solids retention time (days); X_1 = concentration of biological solids in the reactor (mg/ℓ); $\hat{\mu}$ = maximum specific growth rate (mg/mg day); K_s = Monod concentration constant (mg/ℓ); (i.e., substrate concentration when specific microorganism growth rate equals one half of the maximum rate. k_d = specific microorganism decay rate); (mg/mg day), Y = microorganism yield on substrate Plots of the kinetic coefficients $\hat{\mu}$, K_s and k_d vs. the cation concentration were constructed. Design charts relating substrate removal efficiency and BSRT for various cation concentrations enabled application of the experimental results to a more general range of design situations. The complex composition of most wastewaters and the diversity of metabolic pathways of the various groups of symbiotic bacteria forming an anaerobic fermentation exacerbate the task of delineating the effects of antagonism, synergism, complex formation, and acclimation on the behavior of inhibitors such as light metal cations. A large amount of detailed information from the

Table 3

**COMPARISON OF THE TYPICAL ALKALI METAL ION
CONCENTRATIONS FOUND IN DAIRY MANURE WITH THE
RECOMMENDED MAXIMUM ALKALI METAL ION
CONCENTRATIONS ALLOWED BEFORE THE ONSET OF
INHIBITION OF ANAEROBIC DIGESTION**

| | | Recommended maximum concentrations for anaerobic digestion[b] | |
Cation	Concentrations in fresh manure[a] (molar)	No other cation present (molar)	With antagonists present (molar)
Na	0.003—0.120	0.3	0.35
K	0.022—0.145	0.13	0.35
Ca	0.019—0.086	0.15	0.20
Mg	0.017—0.060	0.065	0.14

[a] Values calculated from data of Safley et al.[83]
[b] Values recommended by Kugelman and Chin.[77]

digestion of a variety of wastewaters is required to refine an overall kinetic model applicable to methane digesters.

High concentrations of the alkali metals Na, K, Ca, and Mg are not common in livestock wastes.[11] However, in some instances high levels of NaCl are used in beef feeding rations. Shuyler et al.[82] compared the sodium content in the manure of beef cattle having 0.5, 0.25, and 0.0% NaCl added to their feed ration. Sodium levels in the manure of the 0.5% NaCl group were up to 333% higher than in the group with no added sodium. Table 3 compares the range of alkali metal ion concentrations in dairy cow manure measured in a 1 year study by Safley et al.[83] with the levels recommended by Kugelman and Chin[77] to maintain anaerobic digestion at ≥ 90% of the rate expected without inhibition.

3. Heavy Metal Inhibition

Trace quantities of certain inorganic elements including heavy metals such as zinc, iron, cobalt, copper and manganese are required for optimum bacterial growth.[46] Many of these elements are cofactors necessary for microbial enzymatic reactions, but at greater than trace levels, heavy metals have been found to be extremely toxic to biological waste treatment systems. Inhibitory concentrations reported in the literature vary widely.[77] Variability is probably due to the number of environmental factors that effect the toxic influence of heavy metals. These factors include: precipitation by sulfides due to complex formation, digester pH, temperature, acclimation, detention times, antagonism, and synergism.[84] Laboratory studies by Mosey and Hughes[85] ranked the relative toxicities of heavy metals as follows:

$$Zn = Cu = Cd > Cr(VI) = Cr(III) >> Fe$$

Iron is much less toxic than the other heavy metals. Fermentors can tolerate relatively high levels of iron since ferrous hydroxide is precipitated at normal reactor pH.

Lawrence and McCarty[86] noted that heavy metal toxicity could be eliminated by the addition of approximately 0.5 mg of sulfide per 1 mg of heavy metal causing heavy metal precipitation as sulfide salts. Kugelman and Chin[77] concluded that ferrous sulfate was the most convenient form of sulfide addition. Sulfides would be produced by the biological reduction of sulfate while the excess sulfide would be held out of solution by complexing with iron. Other heavy metals entering the digester will preferentially replace the iron in the complexed sulfide salts because iron sulfide is the most soluble of heavy metal sulfides.

Kugelman and Chin[77] found that in the presence of heavy metals, gas production decreased much more rapidly than the rate of volatile acids accumulation indicating that heavy metals were equally as toxic to the acetogenic bacteria as they were to the methanogens.

Heavy metals are not usually a problem in livestock waste fermentors since these metals are present in only trace quantities in animal manures. As well, the sulfides derived from proteolysis during manure fermentation are generally sufficient to precipitate toxic levels of heavy metals. However, problems could arise in some swine operations that employ copper in the feed ration as a chemotheraputic agent. This increased level of Cu in the feed can cause problems with anaerobic treatment systems.[11]

4. Volatile Acids Inhibition

The methanogenic bacteria exhibit a slower growth rate and the greatest sensitivity to variations in environmental conditions. The volatile acids produced by the acetogenic bacteria are the primary metabolic substrates of the methanogens. It has been demonstrated that high concentrations of volatile fatty acids[14] or pH levels below 6.4[87] are inhibitory to the growth of the methanogens. An empirical rule of thumb states that the normal levels of volatile acids in an anaerobic digester should not exceed 500 mg/ℓ as acetic acid.[84] Acetic acid typically accounts for approximately 70% of the short chain fatty acids produced during anaerobic digestion. Propionic was the next major acid while formic, butyric and valeric acids constituted a small fraction.[3,88]

Various researchers have found different levels of volatile fatty acids causing inhibition during methane fermentations. These conflicting reports can possibly be explained by the effects of pH. It has been found that the concentration of unionized volatile acids was mainly responsible for methanogenic inhibition.[34,47,84] It has been demonstrated that digesters can operate at higher than normal volatile acid concentrations if the pH is maintained high enough to hold the concentration of the unionized volatile acids to less than 30 mg/ℓ as acetic acid.[84] McCarty et al.[88] have shown that slug loadings of various combinations of volatile acids totaling 8000 mg/ℓ as acetic acid were not inhibitory to methanogenesis in digesters with controlled pH. McCarty et al.[88] did find however that relatively high levels of propionic acid were inhibitory to the acid forming bacteria. This may have been the result of H_2 inhibition (Section IV.C.7) rather than inhibition by propionic acid. Duarte[89] suggested that gas chromatographic analysis of volatile acids could provide useful information determining the cause of inhibition in a field situation since a rising propionic acid concentration is a good indication that the acetogenic bacteria have been inhibited.

5. pH Inhibition

An anaerobic digester pH of less than 6.0 or greater than 8.0 rapidly inhibits methanogenesis under most conditions and will lead to digester failure regardless of the volatile acids concentration.[87] However, pH levels outside the 6.0 to 8.0 range are not bactericidal. Keefer and Urtes[90] demonstrated that methanogens could survive for long periods at a pH < 5.0 with a lag period for recovery after pH was returned to normal. Clark and Speece[91] demonstrated recovery after inhibition above pH 8.2 with less of a lag period than recovery from low pH. Variations in pH are very likely to occur during digestion of wastes that are low in nitrogen or that are strongly acidic or alkaline. A COD:N ratio greater than 100:2 cannot supply enough ammonia to support the formation of the buffer ammonium bicarbonate. In addition, low nitrogen levels may result in growth inhibition of the methanogenic bacteria since the rapidly growing acetogens assimilate nitrogen much more efficiently than the methanogens.[84] Neither of these two situations occur often with the anaerobic digestion of livestock wastes. Low pH in farm digesters seldom occurs as a result of low nitrogen levels.

Stanier et al.[92] proposed an hypothesis for the significance of pH and the level of dissociation of volatile acids and ammonia on bacterial inhibition. Weak acids and bases are poorly dissociated at low and high pH values. The undissociated forms of acids and bases easily migrate through cell membranes and thus can change the internal cell pH, but dissociated forms of acids and bases do not enter cells so easily. Thus weak acids and bases exert a toxic effect on cells at low and high pH respectively.

6. Ammonia Inhibition

Ammonia has also been shown to produce an inhibitory effect on methanogenesis in anaerobic digesters. Ammonia inhibition has also been found to be pH related. The relationship between pH and ammonia toxicity was observed by McCarty and McKinney.[93] Sathananthan[94] demonstrated that it was the free ammonia (NH_3) concentration and not the ammonium (NH_4^+) ion concentration that was inhibitory. He found that a free ammonia concentration of >80 mg/ℓ would begin to cause inhibition. The ratio of $NH_3:NH_4^+$ in solution increases with increasing pH. The pH dependent concentration of ammonia is expressed by the ionization equation.

$$K_a = [H^+][NH_3]/[NH_4^+] \tag{7}$$

Where: K_a = the ionization constant. The value of K_a at 35°C is 1.13×10^{-9}.[93] Wastewaters high in organic nitrogen produce high levels of ammonia during anaerobic digestion, but according to the value of the ionization constant, the total ammonia concentration at pH 7.5 to produce 80 mg/ℓ as free ammonia would have to be approximately 7000 mg/ℓ. Such high ammonia concentrations would not be often expected during anaerobic digestion of animal manure. Rather than create a toxic condition, the levels of ammonia typical in the anaerobic digestion of livestock waste contribute to digester stability by contributing to the buffering capacity by the formation of ammonium bicarbonate.[47,95] This naturally produced buffering capacity is measured in terms of "alkalinity". An acid titration to pH 3.7, expressed as equivalent milligrams per liter of $CaCO_3$, has been recommended by the American Public Health Association[96] as a means of measuring alkalinity in an anaerobic digester. Carbon dioxide is also a byproduct of the fermentation of organic wastes.[97] Since there is a high partial pressure of CO_2 in the digester, much of it remains in solution as HCO_3^- and forms NH_4HCO_3 acting as a buffer.[47] It has been suggested that alkalinity in the range of 2500 to 5000 mg/ℓ as $CaCO_3$ is an indication of a stable digester.[24]

Volatile acids react with ammonium bicarbonate to form volatile acid salts:

$$CH_3COOH + NH_4HCO_3 \rightarrow CH_3COONH_4 + H_2O + CO_2 \tag{8}$$

Up to 80% of the alkalinity (buffering capacity) can be reacted with the volatile acids without concern;[20] beyond this, there is a danger that all the ammonium bicarbonate will be exhausted and the free acid will lower the digester pH. A sudden change in the closely monitored level of volatile acids is a better indication of impending digester problems than is an isolated measurement of volatile acid concentrations.[20,46]

7. H₂ Inhibition

Methane formation during anaerobic digestion by CO_2 reduction:

$$CO_2 + 4H_2 \rightarrow 2H_2O + CH_4 \tag{9}$$

typcially contributes 30% of the total methane produced, while the remaining 70% is attributable to acetate splitting:[98]

$$CH_3COO^- + H_2O \rightarrow HCO_3^- + CH_4 \tag{10}$$

Acetate splitting was found to be the rate-limiting reaction, thus the under utilized capacity of CO_2 reduction normally maintains a very low partial pressure of H_2 in the reactor.[98] Shea et al.[99] found that the rate of hydrogen removal during normal anaerobic digestion was typically only 3% of its maximum possible rate. The normal H_2 partial pressure during anaerobic digestion is approximately 10^{-4} atm. However, excessive slug loadings, or other unbalanced digestion conditions can lead to H_2 accumulation. Kaspar and Wuhrman[98] found that H_2 partial pressures greater than 0.5 atm were inhibitory to the oxidation of propionate:

$$CH_3CH_2COO^- + 3H_2O \rightarrow CH_3COO^- + HCO_3^- + 3H_2 + H^+ \qquad (11)$$

During normal steady state operation, the amount of propionate oxidized was about 12% of the amount of acetate oxidized. They found that high H_2 levels did not inhibit oxidation of the longer chain fatty acids, so that propionate and H_2 accumulated from the oxidation of these higher fatty acids, causing the pH to fall thus leading to digester failure.

McCarty et al.[88] proposed that propionic acid was inhibitory to the acid forming bacteria in an experiment using laboratory scale anaerobic sewage sludge digesters. This phenomenon may instead have been attributable to H_2 inhibition of the propionate oxidation reaction. The large batch loadings of propionate employed may have produced a sudden surge of H_2 formation, inhibiting the further splitting of propionate to acetate, CO_2 and H_2.

8. Sulfide and Sulfate Inhibition

Sulfide production in anaerobic digesters usually results from the reduction of sulfates, from the degradation of proteins, or from the introduction of sulfide into the raw wastes. Soluble sulfides have an inhibitory effect on methane formation. McCarty[76] noted that soluble sulfides in excess of 200 mg/ℓ, as sulfur, are inhibitory even in an acclimated digester. Sulfides can be found as insoluble salts of heavy metal cations, as soluble sulfides, and as gaseous H_2S. The amount of soluble sulfide that is removed as gaseous H_2S from the digester liquid is a function of the overall gas production rate and the digester pH.[76] High gas production rates and low digester pH will tend to decrease the concentration of soluble sulfides in the digester and increase the levels of gaseous H_2S emitted. Rudolfs and Amberg[100] found that soluble sulfides had little effect on volatile acids formation although they were inhibitory to methane formation.

Gas scrubbing, precipitation with iron salts, or removal of sulfate from the organic feed to the digester are means of lowering toxic concentrations of sulfide.[76]

Nonreduced sulfate can also be inhibitory to methane digestion by a different mechanism. Sulfates have been shown to induce a competitive type of inhibition to methanogenesis.[101,102] When sulfate is added to methane fermentation, it appears that sulfate-reducing bacteria preferentially use the available acetate and H_2 for the reduction of sulfate, thus inhibiting the formation of methane. The competitive advantage of sulfate-reducing bacteria over the methanogens is theoretically substantiated by comparing the thermodynamic energetics of the competing reactions.[103] The reduction of sulfate to sulfide by H_2 yields 14% more energy than the reduction of CO_2 by H_2 to methane, and the reaction:

$$SO_4^{2-} + CH_3COO^- + H^+ \rightarrow H_2S + 2HCO_3 \qquad (12)$$

yields 66% more energy than the acetate splitting reaction:

$$CH_3COOH \rightarrow CH_4 + CO_2 \qquad (13)$$

The result of sulfate competition with methanogenesis is the reduction of methane production. Thus, the recommended treatment for heavy metal toxicity, the addition of ferrous sulfate as previously described, would reduce methane production if the sulfates became present in excessive amounts. Anderson et al.[84] have reported the discovery of a stable anaerobic microbial association that prevented the formation of H_2S during anaerobic digestion of high sulfate wastewaters and maximized the yield of methane in the presence of sulfates to the extent that there was no measurable level of hydrogen sulfide in digesters using this microbial association.

9. Miscellaneous Toxic Materials

McCarty et al.[104] found that solubilized long chain fatty acids, such as palmitic, stearic, and oleic acids, exerted a toxic effect on anaerobic digestion. However, this form of toxicity may be quickly remedied by the addition of calcium chloride to form insoluble calcium salts of these acids. For example, inhibition due to sodium oleate, a common fatty acid in ordinary soaps, was caused by concentrations exceeding 500 mg/ℓ.[76] This type of inhibition was successfully treated by addition of calcium chloride to cause precipitation.

Long chain fatty acids are broken down during anaerobic digestion mainly by the beta-oxidation pathway.[3,105] The beta carbon of a fatty acid is oxidized, giving up H_2, followed by cleavage of an acetate molecule. This process of two carbon cleavage is then repeated with the remainder of the fatty acid. The beta-oxidation of long chain fatty acids appears to be accomplished by a separate group of acetogenic bacteria.[3] This group of acetogens has a relatively slow growth rate compared to other groups of fermentative bacteria in an anaerobic digestion. However, fairly high levels of fatty acids can be digested if a sufficiently long start-up or acclimation period is provided to build-up the population of acetogens that degrade long chain fatty acids.

Slug additions of 50 mg/ℓ of nitrate (NO_3^-) were found to cause methanogenic inhibition.[106] However, a significant population of facultative bacteria will rapidly reduce the nitrate and prevent toxicity.[104]

Various substances such as phenols, methane analogs (such as chloroform), and ABS plastic have been reported as toxic to methane bacteria.[104]

Cowley and Wase[9] have cited examples of several hazardous substances that are typically found in farming situations. These include chlorinated hydrocarbons often present in the detergents used in milking parlors and piggeries, phenolic disinfectants, and antibiotics used to treat animal diseases.

Atkinson and Swilley[107] demonstrated inhibition of an aerobic bacterial fixed-film reactor by partial physical blockage of the bacteria of the attached film. Two types of film shielding were observed, particle entrapment and precipitate adsorption. In one case, suspended solids in the liquid phase were entrapped on the surface of the gelatinous bacterial film; and in the second case ferric chloride appeared to be adsorbed onto the film. The effect of film blanketing was to lower the reaction rate. It was not obvious whether the reduced reaction rate caused by ferric chloride adsorption was attributable to a reduction of the available nutrient transfer area of the bacterial film surface, or to a biological effect which reduced the flux of substrate uptake.

These mechanisms of fixed-film inhibition may also apply to anaerobic systems. No work appears to have been done to determine whether blanketing inhibiton of fixed-films may result from the treatment of toxic inhibition by the precipitation of the toxic substances.

V. INNOVATIONS IN REACTOR DESIGNS

A. Biological Considerations for Biomass Retention

1. Increased Biomass Densities

The minimum size of a CSTR for anaerobic digestion is restricted by the slow growth and substrate conversion rate of anaerobic bacteria. The efficiency of waste conversion in a CSTR is low with an HRT of less than 10 days. Bacterial wash-out typically occurs at less than 5 days HRT.[108] The capital cost per unit volume of an anaerobic digestion system is an inverse function of the digester volume. Methods of attaining short HRT's can significantly reduce the reactor volumes required for a particular application. Shortened HRT's can be achieved by retaining the active bacterial solids in the reactor for periods longer than the HRT. Various methods of biomass retention or recycling have led to recent major advances in anaerobic reactor designs. These methods of biomass retention are often based on the ability of anaerobic bacteria to form attached-films on solid support surfaces or to form flocs. Designs of proven capability have been applied mainly to soluble wastewaters.[41] However, more recently, designs utilizing film attachment to solid surfaces or flocculation as means of biomass retention are being developed to handle highly particulate wastewaters such as farm manure, crop residues, and sewage sludges.

2. Floc Formation

Flocculation of bacteria during an anaerobic digestion occurs naturally to a greater or lesser extent, depending upon the environmental conditions. Pavoni et al.[109] found a correlation between bacterial flocculation and the level of extracellular polymer production. Tenny and Verhoff[110] found that flocculation of microorganisms was facilitated by the reaction of extracellular polymeric substances with the binding proteins associated with transport of cell metabolites. This extracellular polymer is mainly polysaccharide and has been found to have the ability to concentrate metallic ions and nutrients.[111] This phenomenon may partially explain the facility of digesters utilizing fixed-film or flocculated bacteria to achieve efficient substrate conversion of dilute wastewaters. Photomicrographs revealed that attached slimes were composed of a relatively small volume of bacteria dispersed in a large gelatinous matrix of polysaccharides.[112] Histological stains diffused very slowly through the denser layers of polysaccharide adjacent to cell surfaces, suggesting that nutrient diffusional limitations may inhibit growth in thicker flocs and films. The amounts of extracellular polysaccharides formed in suspended microbial cultures was found to increase if the levels of either nitrogen, phosphorous or sulfur approached growth-limiting concentrations.[113] The dry weight of attached slime production also increased as the carbon to nitrogen ratio of the bacterial substrate was increased.[114]

Biomass retention in continuous bioreactors by bacterial floc formation depends upon efficient separation of the floc from the liquid phase. Increased floc settling velocities are promoted by increasing floc sizes, floc densities, and decreasing liquid velocities in the settling zone. Callander and Barford[41] have outlined a number of physical and biochemical factors which promote higher rates of floc settling. Floc sizes tend to increase: under conditions of low turbulence, during conditions of low nutrient to microorganism ratio (typically utilized during reactor start-up), during operational conditions which tend to retain or concentrate flocculated microorganisms, and with the use of certain synthetic polymer flocculating agents. Higher floc densities are promoted by minimizing the filamentous bacterial forms that lead to low density flocs by promoting the uptake of inorganic precipitates by flocs; and by minimizing the levels of gas entrapment in flocs.

3. Attached-Film Formation

Attached bacterial films occur in nature as slime on submerged solid surfaces. These appear as sticky or gelatinous films and are composed of microorganisms dispersed in an extracellular matrix composed mainly of polysaccharides.[113] These are similar to the extracellular matrix occuring with flocculation. In general, increased heterogeneity of bacterial species forming an attached film leads to a healthier, more robust film growth.[115] It has been established that mixed microbial cultures that thrive in wastewaters readily form attached films on anthracite, glass, plastic, sand, and stone.[116] The attachment of the initial layer of cells to a solid surface is due to a variety of cell wall support surface interactions.

Mechanisms responsible for microbial attachement to a support surface were discussed by Kolot.[117] These include: electrostatic interactions between charges on the cell surface and the support surface; ionic bond formation between carboxyl and amino groups on the cell surface and a reactive ligand on the support surface; and partial covalent bond formation between the carboxyl and amino cell surface groups and hydroxyl groups (or in the case of glasses and ceramics the dissociated silinol groups) on the support surface. The strength of initial cell attachments depends not only on the cell wall composition and the support surface properties but also on the pH and ionic strengths of the solutions, the cell age, level of surface charges, and the amount of support area exposed.[118]

Marcipar et al.[119] studying four different microorganisms found that the level of cell adsorption to a specific support surface increased with decreasing pH from pH 6.0 to 4.0. Kolot,[118] reported that ionic strength also affected the level of microbial adsorption. Increased adsorption occurred when the ionic strength of the solution was increased at constant pH.

Characklis[114] proposed that the length of the induction period required to establish the initial monolayer of attached cells is inversely proportional to the number of nucleation sites available on the support surface. Nucleation sites were defined as the preferential places on the support surface that offered either physical or chemical means of cell attachment. These include small surface pits that provide shelter to a cell from liquid shear forces or a locally favorable surface composition offering a possibility for chemical adsorption. Once the initial microbial layer is established, subsequent generations of bacteria grow over the initial layer embedded in a support of gelatinous slime. The growth kinetics of attached bacterial films are thought to occur in three phases.[120] The first phase is characterized by logarithmic growth. The second phase begins when the rate of substrate utilization becomes constant and film growth is linear. The third phase is reached when newly dividing bacteria do not attach to the film.

Liquid velocities can affect bacterial film growths. Heukelekian[121] found that high fluid velocity retards the establishment of the initial bacterial layer on the surface, but once established, the rate of film growth was actually faster. A possible explanation for this phenomenon is that the high velocities provide better nutrient transport from the liquid medium.[122] The final film thickness at the third stage was also effected by the substrate velocity; thinner films occurred at higher velocities.[120]

Experiments by de Vocht et al.[123] have demonstrated that the types of methanogenic associations which occur in an anaerobic reactor are influenced by the different physical mechanisms utilized for biomass retention in various reactor designs. Two reactors were set up to operate with the same substrate and inoculum. One was designed for biomass retention based upon separation by flocculation and settling. The second reactor was designed to retain active biomass by adhesion within porous support particles. The resulting bacterial populations produced in the two different reactor designs were quite different although both had started from the same seed culture.

B. Retained Biomass Reactor Designs
1. Floc Based Designs
a. Anaerobic Contact Process

The anaerobic contact process was the first method of anaerobic digestion to achieve significantly increased concentrations of active biomass. A settling tank was used for the effluent from a conventional CSTR to concentrate the flocculated biomass and undigested solids. The concentrated solids were then recycled to the reactor while the dilute supernatant liquid was passed off as the final dilute effluent. Figure 2 depicts the basic elements of the anaerobic contact reactor design.

In the early 1950s, an anaerobic contact reactor was developed for the treatment of meat packing wastewater.[124,125] BOD removal efficiency of 95% and suspended solids removal of up to 90% were achieved with this reactor. It operated with an HRT of 12 hr at 35°C

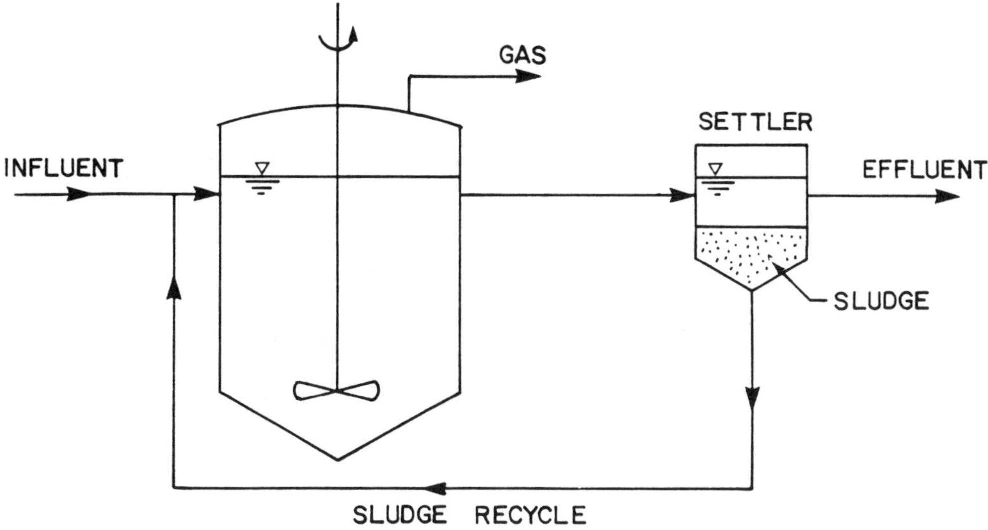

FIGURE 2. Anaerobic contact reactor.

and a loading of 3200 mg BOD/ℓ of digester per day. This reactor design could successfully treat very dilute wastewaters, but operation at 35°C (the optimum mesophilic temperature for anaerobic digestion) was not financially sound since the methane production was not sufficient to heat the relatively large liquid volumes being treated.[126] However, efficient treatment of wastewaters at lower temperatures with the anaerobic contact process can be economically feasible due to the high biomass concentrations and long solids retention times in the reactor which compensate for the effect of reduced reaction rates due to lower temperatures. BOD removal efficiency decreased by only 1.5% when the operating temperature was reduced to 29.5 from 35°C.[125]

The anaerobic contact process has been successfully used to treat various types of wastewaters. van den Berg and Lentz[127] used the anaerobic contact design to digest some difficult-to-treat food processing wastes. These included pear peeling waste, bean blanching waste, potato peeling waste, and rum stillage. Nutrients such as nitrogen, phosphorous, sulfur, and yeast extract were added in order to achieve high loading rates, and, although COD removal efficiencies were up to 90%, several problems were encountered. Solids separation in the settler was often poor. A pilot plant anaerobic contact process was used by Simpson[128] to treat an urban waste similar to domestic sewage with a COD of 1100 to 1300 mg/ℓ at a loading of 640.7 mg BOD/ℓ/day with an HRT of 12 hr. Sludge separation was also a problem with this experiment and a vacuum degasification system was employed to improve the sludge settling characteristics.

The anaerobic contact process is typically operated with digester biomass densities of 5 to 10 g VSS/ℓ with loadings in the range of 2 to 6 kg COD/(m³ day) using low to medium strength wastes (approximately 5000 mg COD/ℓ). When applied to high strength wastes (approximately 20,000 to 80,000 mg COD/ℓ), the anaerobic contact process typically is operated with biomass densities of 20 to 30 g VSS/ℓ and loading is in the range of 5 to 10 kg COD/(m³ day).[41]

Solids separation is a recurrent problem with anaerobic contact reactors. This problem is most prominent with treatment of soluble wastes. In this case, the bacteria tend to remain dispersed since there are no suspended solid waste particles to act as nuclei for flocculation. The dispersed microorganisms tend to be lost with the effluent stream.[129] Methods to promote more efficient solids separation to improve upon the conventional gravity settling method

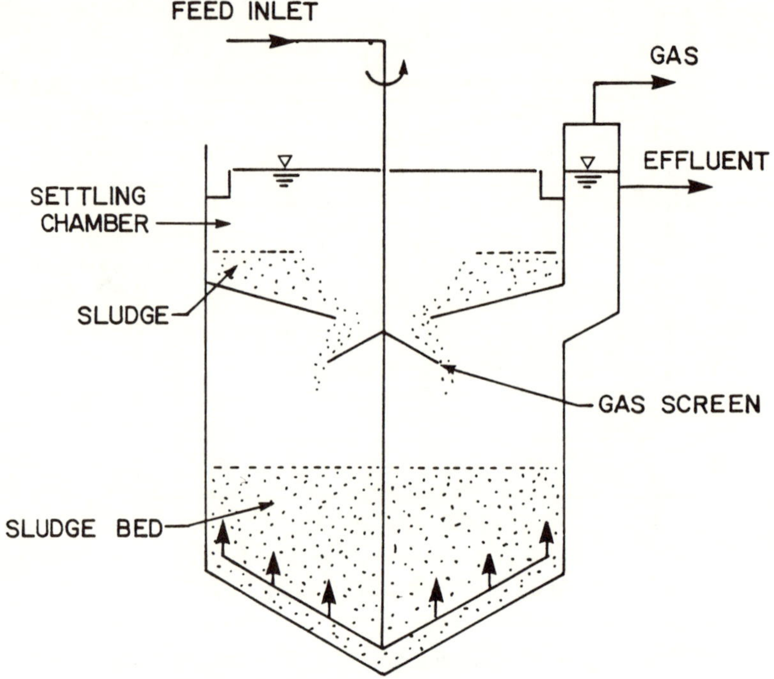

FIGURE 3.   The anaerobic clarigester reactor.

have been employed. These include polymer-enhanced flocs, centrifugation, floatation, vacuum degasifaction, aeration, and thermal shock.[41] The anaerobic contact process exhibited the ability to restart rapidly after a temporary shutdown period.[125] Part of the ability of the anaerobic contact process to achieve increased reaction rates and reduced HRT's is due to the intimate mixing of relatively high levels of biomass solids with the influent wastewater. However, high levels of liquid recycle are required. Sludge recycle to influent flow ratios as high as 5 have been required for efficient COD reductions.[125] As well, the anaerobic contact process is susceptible to washout from high shock loadings.

b. Sludge Blanket Designs

The Dorr-Oliver anaerobic ''clarigester'' was an early sludge blanket reactor design. The clarigester was a variation of the anaerobic contact process that incorporated the settling tank as a compartment at the upper level of the digester.[41] Figure 3 depicts the basic features of this design. The wastewater feed was introduced through inlets spaced around the digester bottom and rose through the sludge bed. Mixing was accomplished by gas recirculation since mechanical mixing was found to be detrimental to biomass flocculation. The top settling chamber allowed flocs to form and settle back to the bottom sludge bed by gravity.[41] This design was mainly applicable to soluble wastewaters, and as in the anaerobic contact process, efficient biomass settling tended to be difficult.

The ''clarigester'' design demonstrated that it could support biomass densities up to 19 g/ℓ and handle loadings of 3000 mg COD/(ℓ/day) of a high strength waste in the range of 20,000 mg COD/ℓ.[130] COD conversions of 97 to 98% were achieved probably due to the partial plug flow nature of waste travel through the reactor.

The upflow anaerobic sludge blanket (UASB) reactor, Figure 4, was a further development of the ''clarigester'' concept based on the ability of anaerobic biomass to form highly settleable particles under favorable conditions. Beds of dense granules with a size range of 1 to 5 mm have been demonstrated after a carefully controlled start-up procedure that allowed

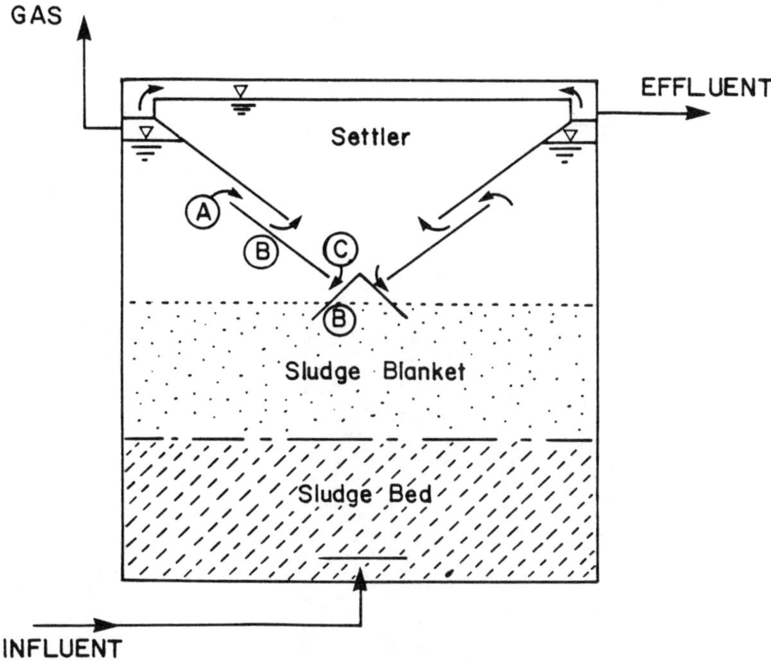

FIGURE 4.   Upflow anaerobic sludge blanket (UASB) reactor. A = sludge/liquid inlet to settler. B = gas screens. C = settled sludge return to sludge blanket. (From Pette, K. C. and Versprille, A. I., Application of the U.S.A.B. — concept for wastewater treatment, in *Anaerobic Digestion — 1981,* Hughes, D. E., et al., Eds., Elsevier Biomedical Press, Amsterdam, 1982, 121. With permission.)

low density seed sludge to wash out of the reactor while biomass with good flocculation and settling properties were retained.[131] The density of the granules collected at the bottom ranged from 60 g TS/ℓ to 10 g TS/ℓ, with flocculent biomass getting less dense toward the gas/liquid interface. Pette and Versprille[132] reported 70 to 90% COD removal with loadings of up to 15,000 mg COD/(ℓ/day) at HRT's as low as 4 hr and sludge concentrations of 40 kg/m^3 in pilot UASB reactors treating beet sugar and potato processing wastes. Full scale UASB reactors have been used for treating wastes from liquid sugar, beet sugar, potato processing, potato starch, brewery, alcohol, and candy plants.

Although the amount of retained biomass per unit reactor volume is enhanced in the UASB design,[133] the development of a bed of highly settleable biomass is not always achievable. Hall et al.[134] found that the UASB design was not successful for the treatment of "thermal sludge conditioning" supernate from a municipal waste facility because of the poor settling characteristics of the biological solids. Heertjes and van den Meer[135] suggested that the UASB could consistently attain greater conversion efficiencies for soluble wastes than expanded and fluidized-bed reactors. This is probably because of the higher biomass retention capacity and partial plug flow characteristics. The UASB design is favored for wastewaters that produce good biomass granules and flocs since mixing requirements are low, biomass retention is high, and the design is less costly than packed bed designs. Callander and Barford[41] report in a review article that the UASB design did not prove feasible for wastewaters with high levels of particulate solids since these tended to clog the sludge bed.

c. Upflow Anaerobic Filters

The upflow anaerobic filter (Figure 5) first devised by Young and McCarty[136] was based on the aerobic trickling filter concept. The filter consisted of a loosely packed bed of 2.5

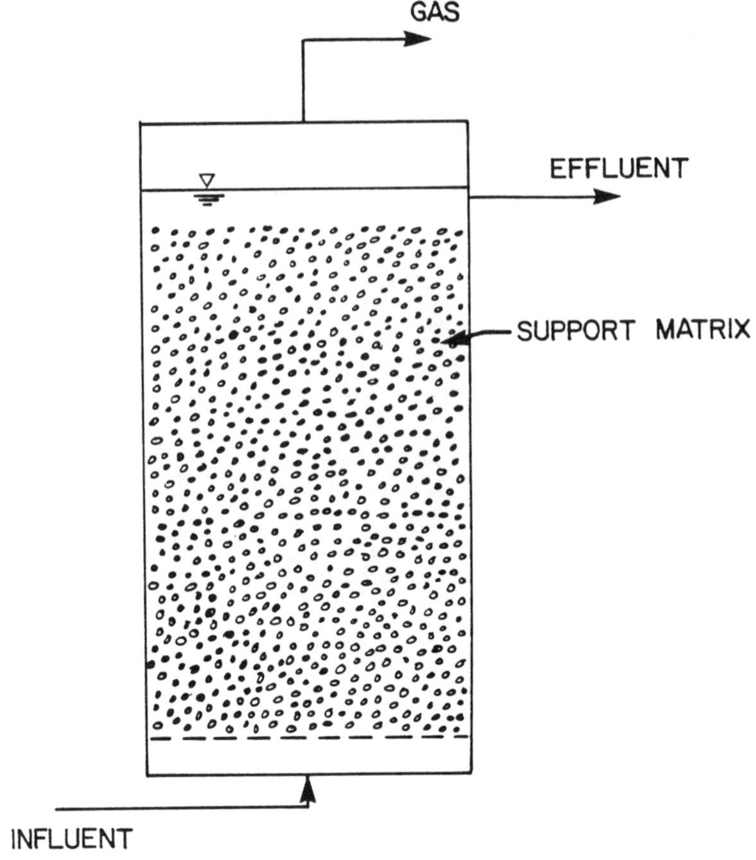

FIGURE 5.   Upflow anaerobic filter.

to 3.8 cm stones. The wastewater was distributed at the bottom of the bed and flowed upward through the interstitial spaces. Biomass retention was facilitated both by film attachment and flocculation. Young and McCarty[136] observed that unattached bacterial flocs were lifted by gas bubbles until striking an overlying stone. When the gas bubbles detach, the flocs settle back into their former position in the void space. This imparts a rolling motion to the flocs which developed spherical shapes that reached diameters up to 0.31 cm. Microscopic examination of these sludge granules revealed the prominence of filamentous anaerobic bacterial shapes that formed intertwined filaments several hundreds of microns in length. These filaments probably influenced the rapid settling of the biological solids. Less of the interstitial flocs were formed in the upper levels of the packed bed, especially when relatively dilute wastes were being treated. Bacteria in the upper layers of the bed were primarily in the form of a film attached to the support surfaces.

Callander and Barford[41] reviewed the performance figures for anaerobic filter designs. Early reactors containing stone packings with voids ratios of about 0.4 achieved biomass densities of 10 to 25 g VSS/ℓ with loadings of 1000 to 6000 mg COD/ℓ. Later designs applied to high strength soluble wastes and employing bed packings with voids ratios of 0.8 to 0.9 have achieved up to 70% COD conversions with loadings up to 10,000 to 20,000 mg COD/(ℓ/day). Colleran et al.[30] reported COD removal efficiencies greater than 80% in an anerobic filter treating agricultural silage runoff with an influent COD of 24,000 mg/ℓ and using an HRT of 3 days at a loading rate of 7,950 mg COD/(ℓ/day).

The anaerobic filter design displayed many of the typical advantages attributable to other reactor designs with increased biomass retention. Low biomass to gas production ratios can be achieved with the anaerobic filter design because of the long solids retention time. Young and McCarty[136] also found that due to the retention of high levels of active biomass, the anaerobic filter could efficiently treat dilute wastewaters at lower than normal mesophilic temperatures. They successfully treated a dilute wastewater at 25°C. This factor is quite significant to the economical application of anaerobic digestion to dilute wastes. Young and McCarty[136] compared several start-up procedures for anaerobic filters. They recommended an inoculation procedure that applied a high seed concentration to the bottom of the filter. This procedure resulted in rapid biomass flocculation and retention in the filter. Light seeding procedures were not as effective. The slow growth rate of methanogens resulted in low levels of unflocculated microorganism that were easily washed out of the filter. Start-up time with the light seeding procedure was long and presented a high risk of reduced filter pH because of a low methanogen population. The anaerobic filter, once established, has proven stable to shock loadings and to influent pH fluctuations.[137,138]

Like the anaerobic contact process and UASB designs, the anaerobic filter is not applicable to treatment of wastewaters with levels of suspended solids is excess of 1%. These designs are therefore not feasible for direct application to the treatment of farm manure slurries with high levels of suspended solids. This problem is exaggerated by the previously cited slow rate of hydrolysis and liquifaction encountered with manure slurry particles. However, Colleran et al.[30] successfully employed an anaerobic filter for the digestion of an hydrolyzed pig slurry supernate as the second stage of a two stage anaerobic digestion. A raw pig slurry was first clarified by gravity settling and hydrolysis during a 2 to 3 week holding period in a storage tank. The clarified slurry from the first stage had a COD content ranging from 10,000 to 60,000 mg COD/ℓ with a suspended solids content less than 1%. The high buffering capacity of the pig manure maintained the pH $\geq$ 6.8. The second stage methanation by an anaerobic filter produced up to 69% removal of COD and 76% removal of the volatile acids at a 3 day HRT with an average loading of 4800 mg COD/ℓ. A series of these studies demonstrated that COD conversions and methane yields were equal to or greater than typical results from conventional anaerobic digesters utilizing a 10 to 25 day HRT. Colleran et al.[30] also reported that optimization of the slurry liquifaction stage could yield methane yields well exceeding those obtained in conventional systems. This two stage concept could be utilized to advantage for the treatment of various livestock manure slurries with other retained-biomass anaerobic reactor designs.

2. Fixed-Film Reactor Designs
a. Stationary Fixed-Film Reactors

High levels of biomass retention in stationary fixed-film reactors (SFFR) are based on the capability of anaerobic bacteria to attach themselves as films to immersed solid surfaces. The SFFR design has been used successfully with low and high density wastewaters including piggery wastes.[139-143] In the SFFR waste circulates vertically up or down through channels which serve as biomass support surfaces. The upflow and downflow patterns produce different operational characteristics. Figure 6 depicts the basic elements of a downflow SFFR.

In the upflow mode, the reactor has the combined characteristics of fixed-film and floc bed reactors.[142] In addition to the formation of an attached film on the support surfaces, flocs and suspended solid particles were retained in the liquid phase between the channels as a partially fluidized sludge blanket. This upflow mode of operation could be useful for providing increased detention times for slowly hydrolyzing solid particles but presents the risk of channel plugging. In the downflow mode, biomass retention was mainly in the form of an attached-film. Channeling and plugging were eliminated in this design since suspended flocs were washed out of the channels. Sludge and solids were drawn off with the effluent at the bottom. The downflow mode behaved as a partial plug flow process (Section VI.A.4.)

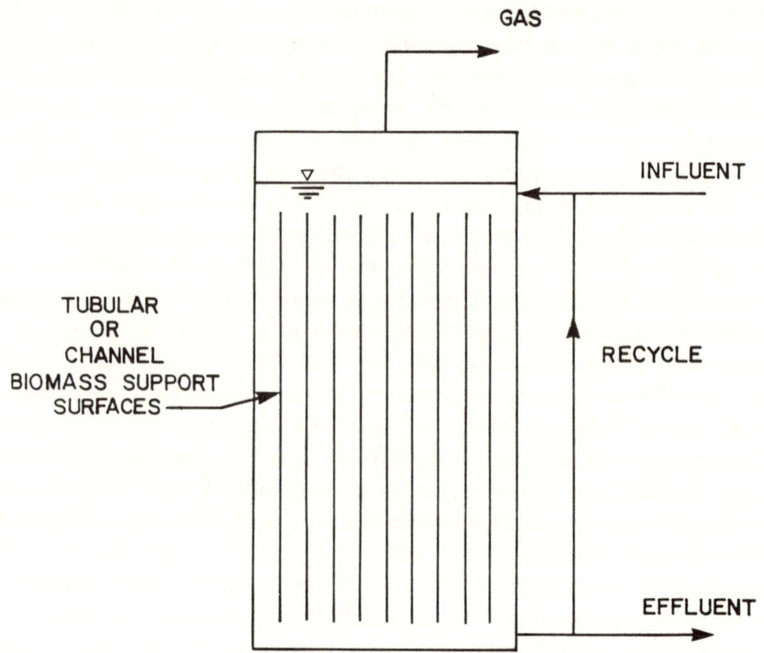

FIGURE 6.   Downflow stationary fixed-film reactor.

capable of the typically high COD conversion rates of plug flow kinetics. The stationary fixed-film system does not require the relatively high energy inputs required by the fluidized-bed reactor designs (Section V.B.2.b) for bed expansion. However, biomass densities in the stationary fixed-film reactor design are limited by the attainable ratios of film support surface area to digester volume. These ratios are in general lower than the ratios attainable with the expanded or fluidized bed designs.[144]

The selection of support material has a significant effect on the length of time required for reactor start-up. van den Berg and Lentz[143] demonstrated that baked clay surfaces required 30 to 90 days for start-up compared to as much as 400 days for glass surfaces. Materials with rough surfaces produce faster start-up and lower film slough-off rates than smooth surfaced supports. However, the final maximum rates of methane production are only slightly affected by differences in the support materials used. Support tubes and channels with heights from 0.60 to 1.11 m and with diameters from 20 to 100 mm have been employed yielding surface areas of 50 to 250 m^2/m^3 reactor volume.[41] van den Berg and Kennedy[140] found that the anaerobic downflow SFFR produced from 2 to 6 m^3 $CH_4/(m^3$ of reactor void volume per day) from substrates such as whey, barley, rum stillage, piggery wastes, chemical industry and cannery wastes, and sewage wastewaters with COD concentrations ranging from 4000 to 130,000 mg COD/ℓ. Kennedy and van den Berg[139] used this reactor design to treat piggery waste with 2 to 4% total solids and 40,000 to 60,000 mg COD/ℓ at 35°C. With a loading rate of 10,100 mg COD/$(m^3$/day), the COD conversion efficiency was 51% at an HRT of 5.2 days, and 41% with an HRT at 2.6 days. The minimum HRT was limited by the relatively slow rate of liquification of organic solids in the pig manure.

Interestingly, van den Berg and Kennedy[141] found that slug loadings at 12 or 24 hr intervals to a downflow SFFR, resulted in higher CH_4 and H_2 production rates than a continuous loading schedule. The opposite would generally be true for UASB or anaerobic filter designs.

Overall, the downflow SFFR displayed the ability to maintain high rates of methane production under adverse conditions. These reactors can withstand low temperatures, shock loadings, sudden waste composition changes, and starvation with the ability for rapid restart.[140]

b. Expanded and Fluidized Beds (Surface Films)

Biological fluidized-bed reactors were developed as an adaptation of the fixed-film concept from the trickling filter reactor design. Weber et al.[145] developed some of the early fluidized-bed bioreactors which used packed and expanded beds of carbon particles to support attached microbial growths for wastewater denitrification. Atkinson and Davies[146] began the development of a kinetic model of fluidized-bed bioreactors. Jeris et al.[147] also utilized the fluidized-bed concept to demonstrate biological dentrification of wastewaters. Fluidized-bed reactors have since been used for the treatment of a variety of low and high strength wastewaters including domestic sewage and relatively dilute (2% TS) dairy manure slurries.[41]

The fluidized-bed reactor combined the advantages of the trickling filter design and the tower fermenter design.[148] Inert solid support particles of a selected size, shape, and density are employed as suspended support surfaces for microbial attachment in fluidized-bed reactors. This design has demonstrated the ability to retain high biomass densities and to provide good biomass/substrate contact without the restrictions imposed by tower fermentors that were limited to low fluid upflow velocities and were very susceptible to bacterial washout due to the low sedimentation velocities typical of microbial flocs.[149] Often loading to tower fermentors was restricted to much lower rates than were required for optimal production because of the danger of washout.

The basic design of the fluidized-bed fixed-film (FBFF) bioreactor is depicted in Figure 7. A bed of solid support particles are maintained as an expanded or fluidized-bed by the frictional forces of a pumped upflow of the wastewater. The direction of liquid flow to support the bed depends upon the relative specific gravities of the solids and liquids. Upward liquid flow is essential for designs with particle densities exceeding the liquid phase density. Conversely, if buoyant forces cause the solids to rise, a downward liquid flow is necessary to disperse the solids. In light of these particle density considerations pointed out by Richardson,[150] a fluidized-bed bioreactor for manure should be designed to operate with a liquid upflow, since manure slurries contain suspended particles that tend to settle, and an upward flow would keep the manure solids dispersed and available for bacterial attack. When the fluidized-bed is in a state of dynamic equilibrium, the net gravitational force (i.e., the force of gravity on the particle minus the buoyant force) and the frictional force of the flow against the particles are balanced. The dynamic pressure drop across the bed from bottom to top equals the weight (less the buoyant force) of the entire bed of fluidized particles.[151] The fluid velocity necessary to keep the particles suspended in a fluidized-bed with a voids ratio approaching one is equal to the settling velocity the particles would attain if they were allowed to freely fall in the quiescent liquid.[150] Voids ratio is defined as the ratio of the void volume between the bed particles to the total fluidized-bed volume. A fluidized-bed can be maintained in equilibrium over a range of liquid flow rates, since the particles in a fluidized-bed with a voids ratio less than one act to effectively reduce the cross-sectional area of liquid flow.[152] Thus, as the voids ratio decreases, the cross-sectional area for liquid flow decreases, and the actual velocity of fluid past the particles is increased for a given mass flow rate through the bed.

The "minimum fluidizing velocity" in a liquid/solid-particle system is the superficial fluid velocity at which the bed changes from fixed to fluidized states.[150] Once the bed is fluidized, pressure drop is constant as flow velocity is increased. The effect of increasing the flow velocity is to increase the voids ratio of the bed.[153] A bed containing uniform particles undergoes a smooth expansion as the fluid velocity is increased from the minimum fluidizing velocity up to the free-falling particle velocity. Nonuniform particles tend to segregate into particle classes.[154] Methods of calculating the required flow velocities for given bed voids ratios have been devised based on a number of easily measured fluidized-bed parameters, such as: free-falling particle velocity, particle size, particle density, particle sphericity, liquid density, and liquid viscosity. Several methods for making these calculations

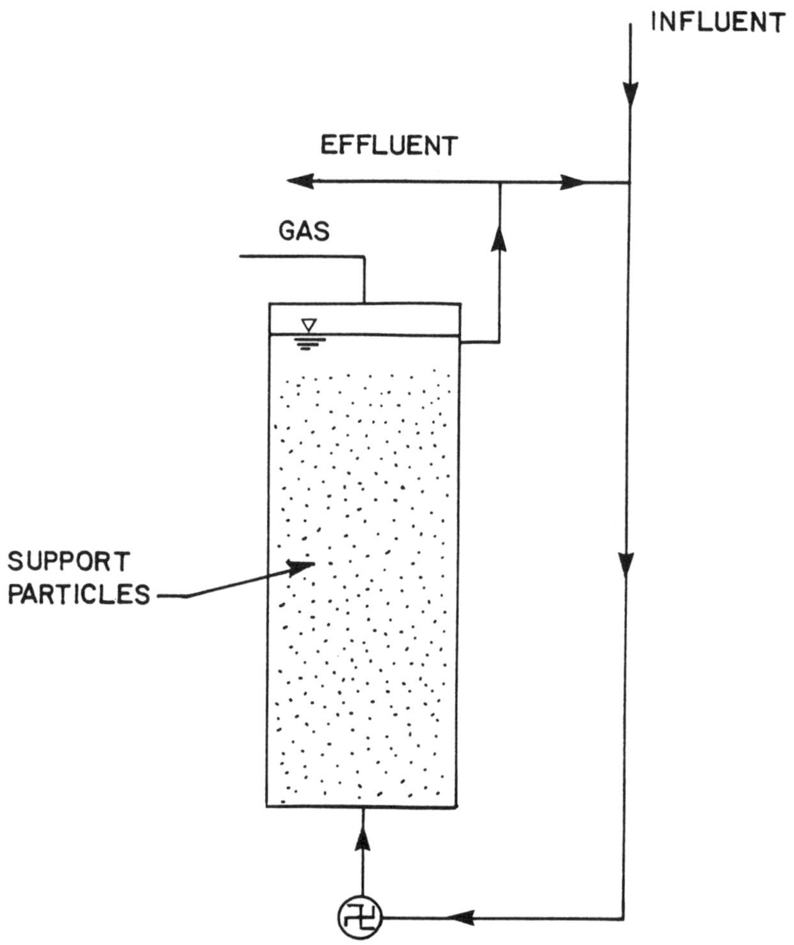

FIGURE 7.   Fluidized-bed, fixed-film (FBFF)reactor.

with minimal experimental data are given by a number of sources. A few of these are: Barnea and Mednick,[155] Ergun,[156] Richardson,[150] Richardson and Meikle,[157] Richardson and Zaki,[151] and Wallis.[158] The viscosity and density of the liquid phase exert a significant effect on the flow rate necessary for a given bed voids ratio, and they are represented either directly or indirectly in equations relating flow rates with bed voids ratios.[157,159,160] The designer of a fluidized-bed system for wastewater treatment is usually not able to adjust the liquid phase characteristics. However, the choice of particles allows one a large degree of control over operating conditions, since the particle density, shape, surface roughness, and volume effect the fluid velocity required for a particular bed voids ratio.[150,161] Sand, carbon, coal, ion exchange resins, PVC beads, porous alumina, aluminum oxide, and glass beads have all been employed as support particles. Larger specific surface areas for biomass support can be attained by reducing the particle sizes. However, the minimum particle size is governed by the tendency for smaller particles to experience elutriation from the bed. Jeris et al.[147] employed 0.65 mm carbon particles and 1.0 mm sand particles in two different fluidized-beds for wastewater denitrification; 0.5 mm sand was utilized in the Dorr-Oliver's Oxitron fluidized-bed system for the aerobic treatment of municipal and industrial wastewaters;[162] Jewell and Cummings,[163] and Jewell et al.[164] utilized anthracite coal, PVC particles, and waste anion and cation exchange particles between 0.5 and 1 mm diameters to treat synthetic wastewater and primary settled sewage both aerobically and anaerobically in fluidized-beds. The wide range of choice for particle supports and for the degree of bed expansion gave very flexible design options.[165]

Atkinson[166] reported several significant advantages of the FBFF bioreactor design. The large biomass holdup potential greatly reduced the reactor size required for a particular application. Furthermore, a steady level of biomass holdup could be achieved by balancing the rate of biofilm attrition due to fluid shear forces and particle collisions against the microbial growth rate. This balance of attrition and growth also meant that the film thickness could be controlled to maintain an optimal bioconversion rate based on the nutrient mass transfer rates from the liquid phase to the film and the diffusional limitations within the film. Atkinson et al.[149] cited an experiment in which a fluidized-bed bioreactor was operated at a constant level of performance for more than a year due to the dynamic balance between biomass growth, decay, and attrition. Atkinson et al.[149] also postulated that fluidized-beds could be operated to regularly strip biomass from the reactor and keep it separated from the effluent water. In this instance, the bed would be fluidized at a velocity low enough to allow a steady increase of film thickness on small dense particles. Accumulation of biomass on the particles reduces their density so that the fluidized-bed will classify them, moving the larger particles to the top of the bed and the smaller particles to the bottom. When the particles become large enough, they will washout of the bed. At this point, they can be sent to a biomass stripper and then be returned to the bed as clean particles. This method of operation can also be maintained at a state of dynamic equilibrium to achieve a constant conversion rate.

The degree of fluidized-bed expansion ranges from 30 to 100%.[41] This degree of expansion requires high fluid flow-through velocities utilizing relatively high pumping power inputs, and high levels of liquid recirculation.[146] To reduce fluidization power requirements, an ''expanded-bed'' design was adopted that typically employed 10 to 20% bed expansions.[167] To further reduce power inputs, particles with lower density were employed. However, reduction in density is limited by the separation efficiency of the particle settling velocities.

FBFF bioreactors have obtained 70 to 90% COD removal from medium and high-strength soluble wastes with loadings of 20 to 30 kg COD/(m^3 day).[168] Anaerobic expanded-bed fixed-film reactors have achieved 60 to 70% COD conversion of low strength (400 to 600 mg COD/ℓ) synthetic sewage at loadings of 10 to 20 kg COD/(m^3 day) with a 30 to 60 min HRT,[169] and 85% COD removal of a high strength soluble dairy whey waste, also at a loading of 10 to 20 kg COD/(m^3 day).[170] Also, relatively highly particulate wastes (2% TS dairy manure) have been successfully digested in an anaerobic expanded-bed fixed-film reactor.[170] During cellulose digestion experiments with this reactor, slowly degraded particulate matter was found to be retained along with the active biomass for periods exceeding the HRT, thus increasing the level of anaerobic liquifaction of the slow to hydrolyse solid particles.

The ultimate thickness of the bacterial film depends upon fluid shear forces over the film surface, physical impact and abrasion, substrate concentration in the fluid phase along with other factors. Diffusional limitations within a microbial film must be considered when adopting the kinetic models for microbial growth and substrate conversion. The rate of substrate conversion per unit film surface area increased with the thickness of the film up to a limit in a study by Atkinson et al.[171] Beyond this, additional film thickness did not improve the reaction rate, since the underlying layer of bacteria was subjected to diffusional limitations. The volumetric reaction rate reached a maximum plateau with high biomass retentions and high volumetric flow rates. The effects of diffusional limitations to substrate uptake have been studied[146,171-175] and a Monod-type[176] kinetic model developed. This kinetic model for attached film reaction rates will be more fully discussed in a later section.

c. The Use of Porous Biomass Support Particles

The utilization of porous particles in a fluidized bed has been a further development of the concept of attached bacterial films for methane production. An infinite number of sizes and shapes may be chosen in order to produce a desired effect.[177] The use of "cell support systems" for immobilized biomass reactors has been reviewed recently by Atkinson et al.[149] In this situation, the biomass is entrapped within the particle instead of growing as a film on the particle surface. The above authors suggested that the particles should have a high internal voids ratio with an open structure. A number of materials have been successfully used as support particles. Stainless steel wire pressed into balls; toroids woven with polypropylene strands; reticulated polyester foams; and matted, reticulated polypropylene sheets have been used as solid support media.[166] As the highly porous particles become filled with bacterial growth, they tended to approach a density of 1.1. Thus, large particles with low denisty can be easily fluidized with reduced pumping power inputs for expanded and fluidized-beds. Well-developed mathematical models have been established to predict the fluidized behavior of spherical shapes.[150,178]

Cuboid-shaped support particles can be produced with one small dimension to provide a constant predetermined thickness to minimize the effects of diffusional limitations.[165] It has been demonstrated by Atkinson,[166] that the lattice structure of the porous particles provides a low shear environment for the entrapped biomass, and a sufficient level of particle/particle contacts could maintain the biomass within the shape of the support. This permits use of bacteria with weak adhesive qualities and facilitates achievement of a predetermined bed-biomass holdup. Atkinson[166] cited further potential advantages for the use of porous particles for entrapped biomass in fluidized-beds. Gas produced by the microbes could escape the particles without causing biomass losses. The physical characteristics of the system, such as the viscosity of the liquid phase could be maintained since there was little biomass freely suspended. The effluent would also contain low biomass levels. Support particles could be regularly removed from the system so that the biomass could be stripped from the particles and kept separate from the effluent. This would be particularly attractive for processes in which the biomass was the desired end product.

Conversion rates and biomass retentions utilizing porous biomass support particles in fluidized beds are similar to results achieved in the solid particle, surface-film fluidized-beds.[41] The advantages of biomass retention with porous particles suggest that they may be useful in reactor designs other than those incorporating fluidized-beds. Alternate designs may be found that are more suitable to highly particulate wastewaters such as farm manures. One example of this is a design utilizing reticulated nylon fiber cuboids that were suspended in an impeller-agitated CSTR. Cuboids with 94% porosity were used for the anaerobic digestion of 6% TS separated dairy manure.[179]

VI. DIGESTER KINETICS

A. Basic Model for Microbial Growth and Substrate Removal

1. Microbial Growth and Substrate Assimilation

The theories of bacterial growth kinetics for various reactor designs have been discussed and reviewed by a number of authors.[38,39,42-44,180-183] The general theory of microbial growth and substrate utilization kinetics will be briefly reviewed followed by a discussion of the application of this model to anaerobic treatment of wastewaters such as farm manures.

Three basic relationships have been employed to model the kinetics of microbial growth.[38,44,180] These describe:

1.   The ultimate bacterial growth rate (i.e., logarithmic phase growth).
2.   The growth rate as a function of available nutrient concentration.
3.   The yield of microbial cells per mass of substrate utilized or converted.

The bacterial growth rate during the logarithmic phase can be described by the equation:[38,39]

$$\frac{dX}{dt} = \mu X \tag{14}$$

or:

$$\frac{dX}{dt} = (\mu - k_d)\, x \tag{14a}$$

where: dx/dt = microorganism net growth rate per unit reactor volume [mass vol^{-1} time^{-1}]; μ = specific microorganism growth rate [(mass of cells formed) (mass of cells)$^{-1}$ time^{-1}]; X = microorganism concentration [mass vol^{-1}]; and k_d = microorganism decay coefficient [(mass of cells decayed) (mass cells)$^{-1}$ time^{-1}]. Equation 14a includes the effects of endogenous respiration and the death rate of microorganisms by including a microorganism decay coefficient, k_d.

Monod[176] porposed a model for the relationship between cell growth rate and the available substrate concentration. This model was initially devised to describe enzyme reaction kinetics but has proven to be applicable to bacterial growth as well. Monod's model has given the best empirical fit over a wide range of substrate concentrations and has been used by many researchers to model substrate conversion during wastewater treatment.[42-44,180-183] The rate of substrate utilization based on the Monod model is often presented in the following form:

$$\frac{dS_1}{dt} = \frac{k\,X\,S_1}{K_s + S_1} \tag{15}$$

where: dS/dt = rate of substrate utilization per unit of reactor volume [(mass of substrate converted) vol^{-1} time^{-1}]; k = maximum rate of substrate utilization per unit weight of bacteria at high substrate concentration [(mass of substrate utilized) (mass of cells)$^{-1}$ time^{-1}]; S_1 = substrate concentration per unit volume in reactor or in effluent [mass-vol^{-1}]; and K_s = half velocity coefficient; equal to the waste concentration when dS/dt is equal to one half of its maximum rate [mass-vol^{-1}].

The yield of cells produced per unit mass of substrate utilized, Y, was considered to be a constant for a given cell-substrate system.[184] This has since been proven accurate enough for most applications to wastewater treatment kinetics:[39]

$$\frac{dX}{dt} = Y\frac{dS}{dt} \tag{16}$$

or:

$$\frac{dX}{dt} = Y\frac{dS}{dt} - k_d X \tag{16a}$$

Equation 16a includes the effect of bacterial decay rate on the yield of cells from substrate conversion. Combining Equations 14, 15, and 16 (or 14a, 15, and 16a, if the decay coefficient is included) gives an equation for the microorganism growth rate as a function of the kinetic constants and the substrate concentration surrounding the cells:

$$\mu = \frac{YkS_1}{K_s + S_1} \tag{17}$$

or, including the microorganism decay coefficient:

$$\mu = \frac{YkS_1}{K_s + S_1} + k_d \tag{17a}$$

The product of Y and k in Equation 17 is often combined into a single term, $\hat{\mu}$, called the maximum specific (or exponential) microbial growth rate [(mass of cells produced) (mass of cells)$^{-1}$ time^{-1}].

2. Kinetics of a CSTR with a Simple Microbe and Substrate System

The CSTR is the most common reactor design for wastewater treatment. The kinetic models for many of the other open continuous fermentation designs are variations of the kinetic model for the CSTR.[38]

Several assumptions are made in the following kinetic model for CSTR's.[44] The liquid flow rate, Q, into the reactor is assumed constant; mixing of the influent in the reactor is complete and instananeous; the effluent flow rate is also constant and equal to the influent flow rate; there are no cells in the influent stream; and the cell concentration in the reactor and effluent stream is equal. The equations for cell growth and substrate utilization for a CSTR have been derived from mass balances for the open system.[38,39,183] For simplicity, the cell decay coefficients, k_d, is not included in the following derivations.

a. Cell Mass Balance

$$\text{increase} = \text{influent} + \text{growth} - \text{outflow}$$

$$\frac{dX_1}{dt} V = X_0 Q + \mu X_1 V - X_1 Q \tag{18}$$

where: V = reactor volume; X_0 = cell concentration in influent [mass-vol^{-1}]; and Q = flow rate through the reactor [vol-time^{-1}]. There is no general analytical solution for this differential equation for cell mass balance. However, for the special case of steady-state operation, $dX_1/dt = 0$. Also, following the previously listed assumptions, $X_0 = 0$. Equation 18 can be reduced to:

$$0 = \mu X_1 - \theta^{-1} X_1$$

or:

$$\mu = \theta^{-1} \tag{19}$$

where: θ = the hydraulic retention time, i.e., V/Q [time]. Equation 19 states that the specific growth rate is equal to the inverse of the hydraulic retention time.

b. Substrate Mass Balance

$$\text{increase} = \text{input} - \text{consumption} - \text{outflow}$$

$$\frac{dS_1}{dt} V = S_0 Q - \mu \frac{X_1}{Y} V - S_1 Q \tag{20}$$

For the steady-state case, $dS_1/dt = 0$, so that:

$$0 = (S_0 - S_1) \mu - X_1/Y$$

$$X_1 = Y(S_0 - S_1) \tag{21}$$

Also from Equations 21, 20, and 17:

$$S_1 = K_s/(\hat{\mu}\theta - 1) \tag{22}$$

The significance of Equation 22 is that the effluent substrate concentration, S_1, in a steady-state CSTR is independent of the feed substrate concentration. S_1 is a function of the hydraulic retention time and the kinetic constants for the system in question. The cell concentration in the reactor will adjust itself according to the level of feed substrate concentration as indicated in Equation 21 for a given effluent substrate concentration.

If the microorganism decay coefficient, k_d, is included in the above analysis then Equations 21 and 22 are modified respectively to:

$$X_1 = Y(S_0 - S_1)/(1 + k_d\theta) \tag{23}$$

and:

$$S_1 = \frac{K_s(1 + k_d\theta)}{\theta(\hat{\mu} - k_d) - 1} \tag{24}$$

Also, implicit in the assumptions used for the CSTR kinetic analysis is that the biological solids retention time, θ_c, is equal to the hydraulic retention time, θ.

$$\theta = \theta_c \tag{25}$$

The minimum hydraulic retention time for a CSTR occurs when the hydraulic retention time is gradually lowered until the effluent substrate concentration becomes equal to the influent substrate concentration. Below the minimum HRT, the cells are washed-out of the reactor at a rate faster than they can reproduce. By letting $S_1 = S_0$ in Equation 17 or 17a, and assuming a high substrate concentration; i.e., $S_0 \gg K_s$, then:

$$\theta^{min} = (Yk)^{-1} \tag{26}$$

or, including the cell decay coefficient, k_d:

$$\theta^{min} = (Y k - k_d)^{-1} \tag{26a}$$

where: θ^{min} = minimum hydraulic retention time before washout occurs, [time]. The efficiency of substrate utilization has been defined as:[43,44]

$$E = 100(S_0 - S_1)/S_0 \tag{27}$$

where: E = the percentage of influent substrate utilized during the process. A different definition of efficiency is based on the volumetric substrate conversion rate for a CSTR:[183]

$$\rho = \theta^{-1}(S_0 - S_1) \tag{28}$$

where: ρ = volumetric conversion efficiency [(g substrate removed) ℓ^{-1} day^{-1}].

The efficiency of influent substrate conversion as a fraction of the feed substrate concentration, as defined in Equation 27, is increased by increasing the HRT. Whereas, the maximum attainable volumetric substrate conversion rate, ρ, occurs at an HRT that approaches the minimum HRT quite closely and generally is regarded as operating too close to the washout point to provide sufficient process stability for a wastewater treatment system. The typical relationships between HRT, effluent substrate concentration, and conversion effi-

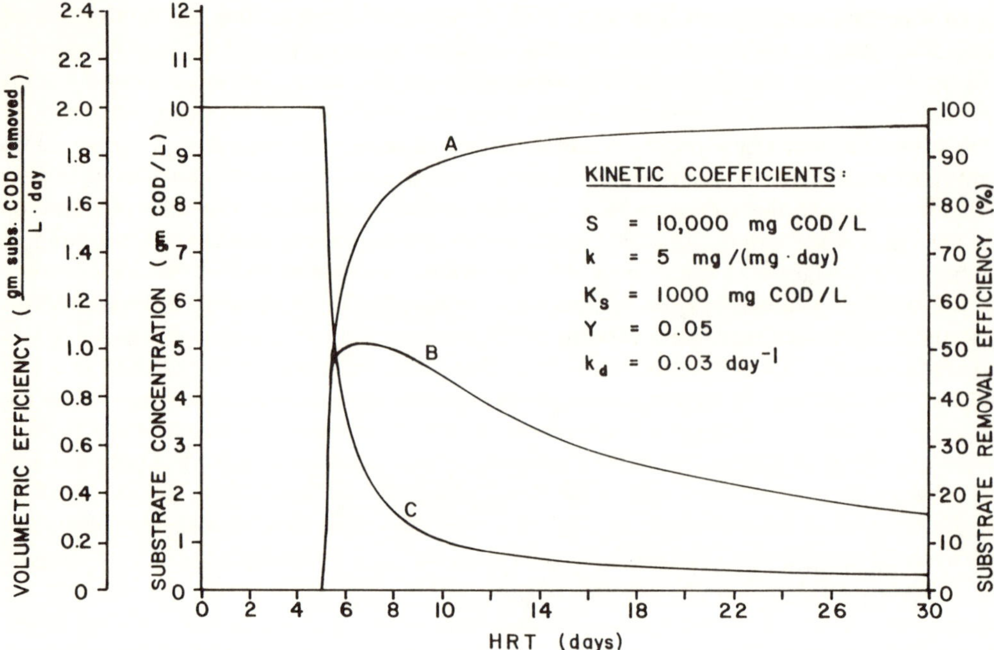

FIGURE 8.   Typical effcciency and substrate removal curves for a CSTR using example kinetic constants applicable to an anaerobic fermentation. A = substrate removal efficiency (Equation 27). B = volumetric efficiency (Equation 28). C = substrate concentration in the reactor (Equation 24).

ciency are shown for an example set of kinetic coefficients in Figure 8. It can be noted in Figure 8 that the peak volumetric substrate conversion rate is quite close to the minimum HRT. As well, the percentage of substrate converted at maximum volumetric conversion efficiency is usually well below the level of influent substrate removal efficiency at longer HRT or BSRT. It was recommended that a kinetic safety factor be applied to the design of an anaerobic reactor.[16] The kinetic safety factor was defined as the ratio of the actual biological solids retention time to the minimum biological solids retention time in a particular continuous fermentation. Typical kinetic safety factors of 3 to 10 have been reported for high rate fermentors.[44] The kinetic coefficients: k, the maximum rate of substrate utilization per unit weight of microorganisms; K_s, the half velocity coefficient for the Monod type kinetic equation; Y, the growth yield coefficient; and k_d, the microorganism decay coefficient have been shown to be constant for a given microorganism-growth/substrate-removal process at a given set of environmental conditions.

3. Kinetics of the Anaerobic Digestion of Complex Wastes

The foregoing kinetic model for a CSTR was developed with the assumption that a single microbial species was being grown on a simple substrate. However, it has been demonstrated that this kinetic model can also be applied to complex multistage microbial conversion processes such as the anaerobic fermentation of municipal sludge or farm manures by employing the concept of a single rate limiting step.[31]

The anaerobic digestion of organic waste materials has been described as a three step process proceeding from the initial liquifaction and hydrolysis of the complex organic material followed by breakdown to volatile acids, H_2 and CO_2 and finally conversion to methane and carbon dioxide.[39,42-44] In most situations, the rate limiting step for this process has been found to be the last stage of the fermentation, i.e., the conversion of the short chain volatile acids to methane.[16,42]

At digestion temperatures less than 35°C, O'Rourke[185] reported that the degradation of long chain fatty acids could be slower than the conversion of short chain volatile acids to methane. However, in most organic wastes such as farm manures the long chain fatty acids concentration is not large enough to significantly change the overall kinetic model based on conversion of short chain volatile acids as the rate limiting step.

Values of the kinetic coefficients, Y, k, K_s, and k_d for the conversion of a short chain volatile acid to methane by anaerobic digestion were measured by a number of researchers.[16,42,185] Lawrence and McCarty[43] found that the values for the growth yield constant, Y, and the cell decay coefficient, k_d, could be considered as constants between the temperatures of 20 to 35°C. Measured values for the yield coefficient, Y, for the conversion of various volatile acids to methane were between 0.040 to 0.054 in units of mg of biological solids produced per mg COD converted to methane. The values of k_d were between 0.010 to 0.040 day^{-1}. The values of k, the maximum rate of substrate utilization per unit weight of active biological solids, for the methane conversion of acetic, propionic, or butyric acids at 35°C were nearly equal if expressed on the basis of mg biological solids produced per mg COD of substrate converted. The COD of substrate converted was based on the equivalent COD of the substrate in the conversion to CH_4. However, the values for k varied with temperature between 20 and 35°C. O'Rourke[185] found values for k for methane conversion of volatile acids from 3.85 day^{-1} at 20°C to 6.67 day^{-1} at 35°C. The value for K_s, the half-velocity coefficient, also varied with fermentation temperature. Lawrence and McCarty[43] and Lawrence[16] found that values for K_s for the fermentation of acetic acid fit the following Arrhenius type equation:

$$\log[(K_s)_2/(K_s)_1] = 6980(^1/T_2 - \,^1/T_1) \tag{29}$$

where: T_2, T_1 = temperatures of fermentation (°K).

Since the value of K_s increases with decreasing fermentation temperature, the biological solids retention time would have to be increased in order to maintain the same effluent substrate concentration and thus, the same efficiency of substrate conversion at reduced temperatures, as seen by Equation 22.

The anaerobic digestion of complex wastes produces a mixture of various long and short chain fatty acids. O'Rourke[185] proposed a method of including the combined concentration of all these volatile acids as the substrate for the kinetic model. A modification of Equation 24 for the prediction of effluent substrate concentration was employed for calculation of the cumulative short and long chain fatty acids concentration. Assuming that the values of the kinetic coefficient Y, k, and k_d are equal for all the fatty acids of concern at a given temperature as had been demonstrated by Lawrence and McCarty[43] and O'Rourke,[185] Equation 24 could be expressed as:

$$(S_1)_T = \frac{K_c(1 + k_d\,\theta)}{\theta(\hat{\mu} - k_d) - 1} \tag{30}$$

where: $(S_1)_T$ = the total concentration of long and short chain fatty acids in the effluent [mg COD-ℓ^{-1}]; and $K_c = \Sigma K_s$ for all the fatty acids concerned [(mg COD of substrate)-ℓ^{-1}]. The substrate concentration in Equation 30 must be expressed on a COD basis in order for the values of Y and k to be equal for the different fatty acids typically found in anaerobic digesters.

Gross nutrient measurements such as total volatile solids, BOD, or COD have also been successfully used as substrate measurements for employment in the kinetic model in the same manner as the simpler specific types of substrates such as volatile acids.[182] BOD has been used as a gross substrate measurement in a number of kinetic studies of anaerobic digestion.[48,186,187] All of the ultimate BOD of a waste may not be biodegradable by anaerobic digestion. Woods and O'Callaghan[42] recommended taking the difference of the initial COD of the feed and the COD of the effluent as a measure of the anaerobically biogradable BOD of the waste at a specified retention time. As the retention time was increased, the value of the BOD measured in this way approached a maximum limit.

Morris et al.[48] studied the kinetics of dairy manure fermentation. They used a similar approach to that recommended by Woods and O'Callaghan[42] to determine the biologically degradable fraction of the volatile solids in cow manure. The biogradable fraction of cow manure was 42.5% of the influent volatile solids concentration. The biodegradable volatile solids concentration was used as the substrate measure in the kinetic model. The kinetic equations yielded predictions that matched the actual lab fermentation trials well. The value of the maximum growth rate coefficient, $\hat{\mu}$, which is equal to the product of the coefficients Y and k, was constant at a value of 1.0 day^{-1} over a range of different feed substrate concentrations at a temperature of 32.5°C. The value of the half-velocity coefficient, K_s, vaired linearly with the substrate concentration according to the following equation:

$$K_s = 1.435(S_0) - 25 \tag{31}$$

where: S_0 = influent substrate concentration (g VS L^{-1}). Morris et al.[48] proposed that the increase in the value of K_s with substrate concentration could have been the result of several factors. At low substrate concentrations, the availability of the substrate may have been limited in the completely mixed dispersed-bacteria reactor by diffusional resistance. Also, the exposure of the surfaces of the organic particulates to enzymatic degradation could have been rate-limiting during this fermentation. Callander and Barford[41] made a similar observation that the rate of hydrolysis of particulate solids was considerably slower in some cases than the formation of volatile acids from soluble materials.

4. Plug Flow Kinetics

The plug-flow continuous reactor design is a fundamental alternative to the CSTR design. An ideal plug-flow bioreactor delivers a continuous stream of substrate influent through the reactor in a pipe flow manner with no longitudinal mixing. The ideal plug-flow reactor has been described as a series of isolated batch cultures moving through the reactor along a time axis.[38] In the ideal case, some form of microbial inoculation of the influent is required, otherwise no growth would occur without longitudinal mixing. In actual practice, a certain amount of longitudinal mixing does occur in a diffusional manner so that incoming fluid elements are inoculated by those ahead of them. Herbert[38] and Lawrence and McCarty[44] have presented the kinetics of plug-flow reactors assuming some level of effluent recycle to the influent stream in order to make simplifying assumptions for an analytical solution of the kinetic equations. Lawrence and McCarty[44] assumed that an external feedback stream was operating to such an extent that the solids retention time was at least five times greater than the hydraulic retention time so that the concentration of cells from inlet to outlet could be taken as constant. The mathematical analysis for a plug-flow reactor that depends on internal microbial feedback by diffusional longitudinal mixing is more complicated.

A plug-flow reactor with some degree of ideal kinetic behavior is more efficient than a comparable CSTR and produces an effluent with lower substrate concentration. As the level of feedback, either external recycle or internal longitudinal mixing, is increased, the kinetic characteristics of a plug-flow reactor approach those of comparably sized CSTR's with similar HRT's. Wehner and Wilhern[188] developed a dispersion model to account for kinetic behavior in the interval between completely mixed and plug-flow. In actual practice, kinetic differences between the plug-flow and CSTR systems are probably not great.[44] The treatment of plug-flow reactors using the kinetic equations for CSTR's would probably be acceptable

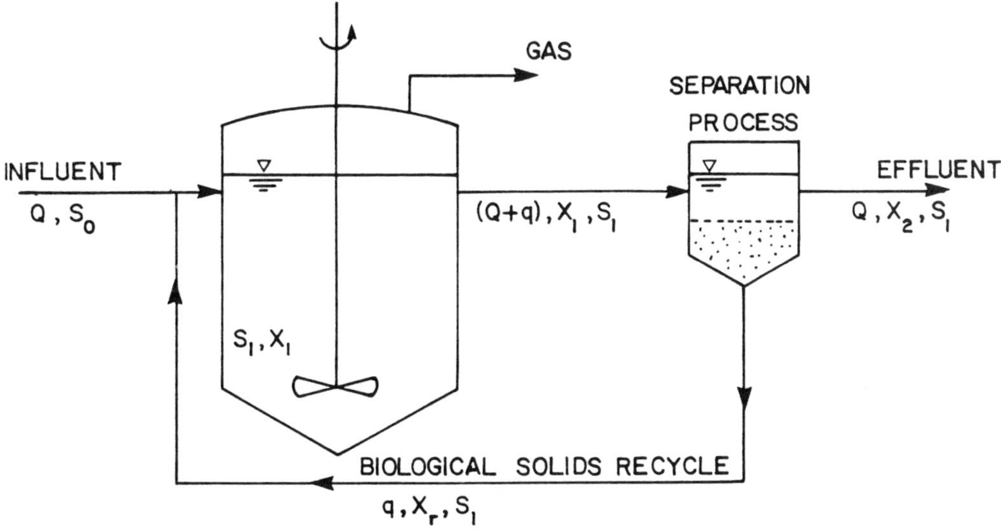

FIGURE 9.   Anaerobic contact reactor indicating substrate and biological solids flow streams.

for most actual design situations. Errors would be on the conservative side leading to better wastewater quality and greater methane production than anticipated.

Plug-flow reactors for methane production from farm wastes have been demonstrated at Cornell University.[7] Their design utilized an excavated earthen trench and involved relatively low capital costs and simple construction. The problem of mixing was also eliminated. The waste conversion and gas production were comparable to conventional CSTR's. However, the reduced capacity of plug-flow reactors to tolerate toxic or shock loadings is a disadvantage to their employment for anaerobic digestion of farm wastes. Toxic or shock loadings are concentrated at one section of the reactor, whereas, in a CSTR, the toxic or shock loading effect is diluted evenly throughout the entire reactor volume.

B. Retained Biomass Reactor Kinetics

1. Anaerobic Contact Process Kinetics

The design and operating characteristics of the anaerobic contact process were described in the previous section. A number of authors have contributed to the development of a kinetic model for the anaerobic contact process. A few of these are Feilden,[181] Herbert,[38] Lawrence,[16] and Lawrence and McCarty.[44] The anaerobic contact process falls under the general kinetic category of a completely mixed biological reactor with solids recycle. The anaerobic contact process was developed to improve upon the substrate utilization rate of the traditional CSTR. It can be seen from Equation 15 for the substrate removal rate for a CSTR:

$$\frac{dS_1}{dt} = \frac{k \, X \, S_1}{K_s + S_1}$$

that if the concentration of cells in the reactor, X, is increased, then the substrate utilization rate also is increased. The anaerobic contact process does this by separating a fraction of cell mass from the reactor overflow and continuously recycling it back to the reactor (Figure 9).

The development of the kinetic model[16,38,44,180] entailed the following two assumptions: (1) substrate utilization occurred entirely in the reactor; and (2) recycle was considered continuous and the volume of the biomass separator was assumed to be negligible so that the total biological mass in the system could be considered to be contained in the reactor. The ratio of the cell mass in the effluent to the cell mass in the reactor can be expressed as:

$$X_2/X_1 = (1 + r - rC) \tag{32}$$

where: $r = q/Q$ volumetric feedback ratio; $C = X_r/X_1$ cell concentration factor; $X_2 =$ cell concentration in the final effluent (mass-vol^{-1}); $X_1 =$ cell concentration in the reactor (mass-vol^{-1}); $q =$ recycle flow rate (vol-time^{-1}); $Q =$ influent flow rate before recycle addition (vol-time^{-1}); and $X_r =$ cell concentration in the recycle stream (mass-vol^{-1}). Using the appropriate kinetics developed in the previous section, a cell balance equation for the reactor is

$$\text{increase} = \text{recycle} + \text{growth} - \text{outflow}$$

$$V \frac{dX}{dt} = r\,Q\,C\,X_1 + \frac{\hat{\mu}X_1\,S_1}{(K_s + S_1)} - (1 + r)\,Q\,X_1 \tag{33}$$

At steady-state operation, $dX/dt = 0$, then, dividing by $(V)\,(X_1)$:

$$\frac{\hat{\mu}\,S_1}{(K_s + S_1)} = (1 + r - r\,C)\,Q/V \tag{34}$$

From Equation 17, the left-hand side of Equation 34 is the specific growth rate, $\hat{\mu}$. As well, the inverse of the specific growth rate is equal to the biological solids retention time, θ_c. Therefore, by substitution in Equation 34:

$$\theta/\theta_c = (1 + r - rC) \tag{35}$$

where: $\theta_c =$ biological solids retention time (BSRT) [days]; and $\theta = V/Q =$ hydraulic retention time (HRT) [days]. From Equations 34 and 35, the effluent substrate concentration of an anaerobic contact reactor is predicted by the following equation:

$$S_1 = K_s/(\theta_c\hat{\mu} - 1) \tag{36}$$

This is similar to Equation 22 for effluent substrate concentration in a conventional CSTR. Including the cell decay coefficient in Equation 36, the substrate concentration in a "contact process" reactor is[16]

$$S_1 = \frac{K_s(1 + k_d\,\theta_c)}{\theta_c(Y\,k - k_d) - 1} \tag{37}$$

Again, using the kinetic equations developed in the previous section, a substrate balance equation for the reactor is

$$\text{increase} = \text{input} + \text{feed-back} - \text{outflow} - \text{consumption}$$

$$V \frac{dS}{dt} = Q\,S_0 + q\,S_1 - (Q + q)\,S_1 - \frac{\hat{\mu}\,X_1\,S_1}{Y(K_s + S_1)}\,V \tag{38}$$

At steady-state, $dS/dt = 0$; then, dividing by (V):

Table 4
SUMMARY OF CHARACTERISTIC EQUATIONS FOR A STEADY-STATE COMPLETELY MIXED CONTINUOUS BIO-REACTOR WITH BIOLOGICAL SOLIDS RECYCLE

Parameter	Steady state kinetic equation	Eq. no.
Effluent concentration	$S_1 = K_s(1 + k_d\,\theta_c)/[\theta_c(Y\,k - k_d) - 1]$	37
Cell concentration in reactor	$X_1 = [Y(S_0 - S_1)/(1 + k_d\,\theta_c)](\theta_c/\theta)$	43
Solids retention time	$\theta_c = [\{(Y\,k\,S_1)/(k_s + S_1)\} - k_d]^{-1}$	44
Minimum solids retention time	$\theta_c^{min} = [\{(Y\,k\,S_0)/(k_s + S_0)\} - k_d]^{-1}$	45
Limiting minimum solids retention time	$(\theta_c^{min})_{lim} = (Y\,k - k_d)^{-1}$	46
Efficiency of substrate utilization	$E = 100(S_0 - S_1)/S_0$	27
Ratio of HRT to biological solids retention time	$\theta/\theta_c = (1 + r - r\,C) = X_2/X_1$	35 and 41
Cell concentration in final effluent	$X_2 = Y(S_0 - S_1)/[1 + k_d\theta_c]$	47

Note: Equations 27, 37, 44 to 46 are the same as those for a CSTR system without recycle except that θ_c has been substituted for θ.

$$X_1 = Y^{-1}\left(\frac{\hat{\mu}\,S_1}{K_s + S_1}\right)^{-1}(Q/V)\,(S_0 - S_1) \tag{39}$$

From Equations 34, 35, and 39; the reactor cell concentration is

$$X_1 = Y(S_0 - S_1)\,(\theta_c/\theta) \tag{40}$$

From Equations 32 and 35:

$$\theta/\theta_c = X_2/X_1 \tag{41}$$

Therefore, the cell concentration in the effluent after separation is

$$X_2 = Y(S_0 - S_1) \tag{42}$$

Table 4 presents a summary of the characteristic equations including the cell decay coefficient for a completely mixed continuous reactor with solids recycle (the anaerobic contact process).

A comparison of cell concentration and cell output for an anaerobic contact reactor with a CSTR without cell recycle is presented in Figure 10. A recycle of one half the influent volume and a twofold cell concentration in the separator were assumed for this figure. The minimum HRT was reduced to one half the minimum HRT without recycle by doubling the biological solids retention time. Cell concentration in the reactor and the cell production are increased by the anaerobic contact process. The relative operating advantages and disadvantages of the anaerobic contact process compared to other retained biomass reactor designs were discussed in an earlier section.

2. Kinetics of Flocculent or Attached-Film Systems

In addition to the external concentration and recycling of biological solids in the anaerobic

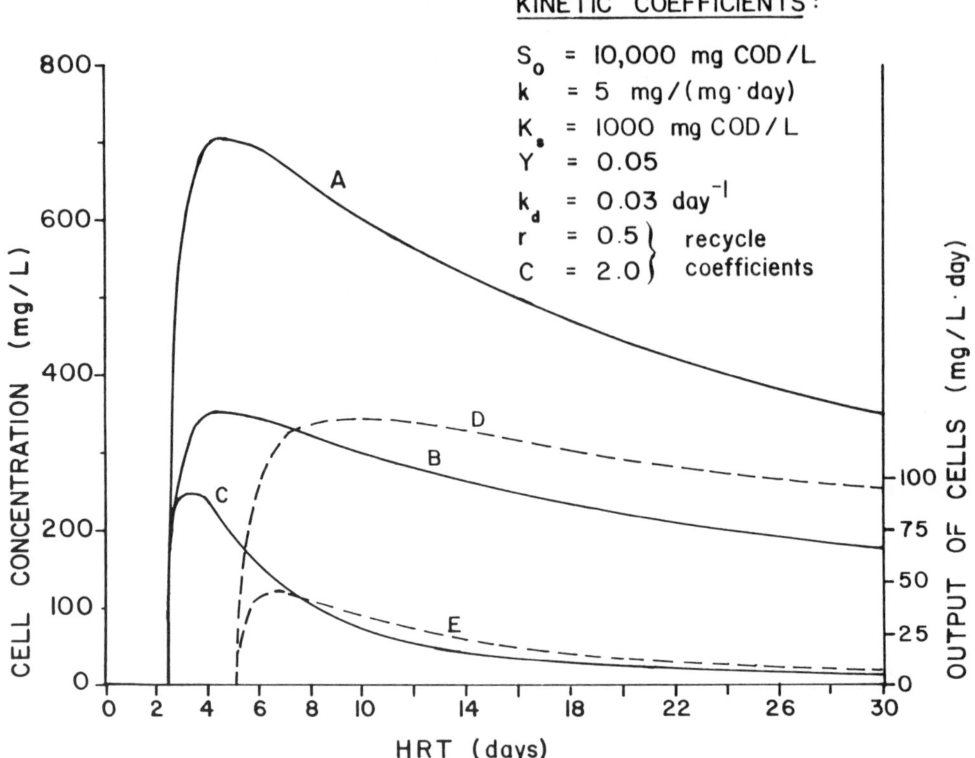

FIGURE 10.   Comparison of cell concentration and cell output for a CSTR and an anaerobic contact reactor using example kinetic constants typical of those applicable to an anaerobic fermentation. A = cell concentration in a reactor with recycle (Equation 43). B = cell concentration in the effluent of a reactor with recycle (Equation 47). C = output of cells from recycle reactor (Equation 47 ÷ HRT). D = cell concentration in reactor and effluent of a CSTR without recycle (Equation 23). E = cell output from CSTR without recycle (Equation 23 ÷ HRT).

contact process, biomass retention in a continuous reactor can be accomplished by internal mechanisms. A number of these systems were described in Section V. These systems separate and retain biological solids through the mechanisms of flocculation, as in the sludge blanket designs; or, attached-film growths on solids support surfaces, as in the fluidized-bed, fixed-film, and the stationary surface fixed-film designs; or, a combination of both mechanisms, as exhibited in the upflow anaerobic filter design.

The cell growth and substrate utilization kinetic relationships for the anaerobic contact process apply in principle to the reactor designs with internal biomass retention, but there are additional factors complicating their application. One of the major differences in growth and substrate utilization kinetics is encountered in the form of reduced nutrient uptake imposed by the builtup layers of microbial growth in flocs or surface attached-films.[174] Other difficulties arise because of the increased complexity of accurately evaluating such basic parameters as the biological solids retention time and the level of active microbial mass within the reactor. The hydraulics of the retained biomass designs are also more difficult to define often lying between completely mixed and plug flow regimes. The UASB and the fluidized-bed reactors are similar in many ways to the plug-flow reactors.

Atkinson and Davies[172] have presented the development of equations for the rate of biochemical reactions in microbial films or flocs. These equations are based on a Monod model for growth rate modified to accommodate the nutrient diffusional limitations inherent in such systems. The rate of substrate utilization by microorganisms per unit surface area of microbial film or floc was expressed in the following form:[172]

$$R_a = \alpha S / (\beta + S) \tag{48}$$

where: R_a = the rate of substrate uptake per unit area of microorganisms; α = a rate coefficient; β = a rate coefficient; and S = the substrate concentration in solution. Equation 48 applies to microorganism growth when it is uninhibited by diffusional limitations. The effect of diffusion on the biochemical reaction rate for the case of a single limiting substrate was developed from the following relationship between diffusion and assimilation:[172]

$$D_e \frac{(d^2S)}{dZ^2} - aR_a = 0 \tag{49}$$

where: D_e = effective diffusion coefficient within the microbial mass; Z = film thickness; and a = external surface area of viable microorganisms per unit volume of microbial mass Equations 48 and 49 were developed into a biological rate equation for microbial films or flocs of the form:[172]

$$N = \lambda N_{max} \frac{K_3 S_s}{1 + K_3 S_s} \tag{50}$$

where: N = rate of substrate consumption per unit interfacial area; λ = effectiveness factor; N_{max} = maximum rate of substrate uptake; K_3 = biological rate equation coefficient; and S_s = substrate concentration at the interface between the microorganisms and the aqueous phase.
Also:

$$\lambda = f(k_1, k_2, k_3, L, S_s)$$

and,

$$N_{max} = K_1 L / K_3$$

where: K_1 = biological rate equation coefficient; K_2 = biological rate equation coefficient; and L = "wet" biological film thickness.

Application of the model to bacterial flocs was accomplished by replacing the characteristic film dimension, L, with a characteristic dimension for flocs defined by the ratio of the floc volume to the floc surface area, V_f/a_f. Equation 50 expresses the overall rate of substrate uptake for a biochemical reaction in a microbial film or floc in terms of a characteristic size, the limiting substrate concentration external to the biomass, and three rate coefficients, K_1, K_2, and K_3. Atkinson and Davies[172] presented techniques for evaluating the kinetic rate coefficients K_1, K_2, and K_3, and for determining the effectiveness factor, λ. Further support and development of this model was reported by Atkinson et al.[146,171,173-175] The effects of diffusional limitations to the rates of substrate utilization by microbial films was examined, and techniques to determine the rate coefficients required to predict the effects of diffusional limitations were devised.

Atkinson[166] emphasized the need for more work on predictive modeling of fixed-film biological reactions to facilitate more efficient designs and to reduce the time and cost for design work.

C. Development of a Dynamic Model

The steady-state kinetic model described in Sections VI, A and B is useful for the initial design of anaerobic digestion facilities. However, this model is restricted. It does not apply to anaerobic bioreactor performance during start-up or other transient conditions of operation. The past record of anaerobic digesters shows lack of process stability with numerous instances of digestion failure.[34] This has discouraged the application of anaerobic digestion for waste-water treatment in spite of its many other significant advantages. The design of a control system to maintain stable operation should be based ultimately on a dynamic model to reflect the inherent dynamic behavior of wastewater treatment systems. Andrews[35] cited a number of benefits that would result from the development and application of a dynamic model. Some of these factors could have great benefits for on-farm applications that require a reliable process capable of efficiently handling fluctuating feed supplies with a means of protection from failure due to shock or toxic loadings.

The Monod type of model for growth and substrate utilization has been most commonly used to model the biological kinetics of an anaerobic digestion process. For the steady-state kinetic model, the basic equation for ultimate bacterial growth rate (Equation 14); substrate utilization rate as a function of available nutrient concentration (Equation 15); and the yield of microbial cells per mass of substrate utilized (Equation 16) as outlined in Section VI,A were employed to establish the differential equations for cell mass balance (Equation 18), and substrate mass balance (Equation 20) for a CSTR. The materials balance equations were then solved for the special case of steady-state operation (i.e., $dS/dt = 0$, and $dX/dt = 0$) to arrive at a set of algebraic equations describing the steady-state operation of a completely mixed continuous bioreactor. For the general case of dynamic operation, there is no readily available algebraic solution to the differential equations for substrate and cell mass balances. Until the introduction of analog and digital computers, the development of dynamic models for anaerobic wastewater treatment systems was not feasible.[35]

Significant advances in the development of a dynamic model of the anaerobic digestion process were reported by Andrews[32,34-35] whose dynamic model was based on material balances of the biological, liquid, and gas phases in a CSTR. The kinetic equations describing cell growth and substrate utilization were coupled with equations to account for mass transfer and accumulation in the liquid and gas phases for such materials as carbon dioxide, bicar-bonate, and cations. To account for the effect of substrate inhibition at high substrate concentration, a modification of the Monod type equation for cell growth rate was proposed by Haldane.[189]

$$\mu = \frac{\hat{\mu}}{1 + \dfrac{K_s}{S_1} + \dfrac{S_1}{K_i}} \tag{51}$$

where: K_i = inhibition constant [mass-vol^{-1}]. The inhibition constant, K_i, is equal to the substrate concentration at one half the maximum specific growth rate that would occur in the absence of inhibition. This equation was further modified to include the effect of pH on growth rate. As noted earlier, the rate limiting step for the anaerobic conversion of complex organic wastes is generally taken to be the conversion of volatile acids to methane. The major volatile acid concerned is acetic. It has been commonly observed that both low pH and high volatile acid concentration were inhibitory to anaerobic digestion. Andrews[34] em-ployed the level of unionized acetic acid as the limiting substrate. This accomodated the inhibitory effect of low pH by expressing the level of unionized acetic acid in terms of the ionization equilibrium equation for dissociation. Equation 51 was modified to the following form:

$$\mu = \frac{\hat{\mu}}{1 + \dfrac{K_s\, K_a}{(S_1)\,(H^+)} + \dfrac{(S_1)\,(H^+)}{K_i\, K_a}} \tag{52}$$

where: K_a = ionization constant for acetic acid; S_1 = total acetic acid concentration (mass-vol^{-1}) ($\cong$ ionized acetic acid concentration at pH $\geq$ 6.0); and H^+ = hydrogen ion concentration (mass-vol^{-1}). The effects of toxic loading were included in the model by assuming a first order equation for the rate of organism death due to toxicity.[190]

$$r_d = K_t T_x \tag{53}$$

where: r_d = rate of organism destruction [mass-vol^{-1} day^{-1}]; K_t = toxicity rate coefficient [(mass of cells killed) (mass of toxic agent)$^{-1}$-day^{-1}]; and T_x = concentration of the toxic agent [mass vol^{-1}].

The rates of carbon dioxide and methane formation were calculated from the growth rate equation and yield coefficients for gas production per unit mass of microorganisms. Other equations were employed in the model for the bicarbonate-carbon dioxide equilibrium in solution, the rate of CO_2 transfer between the liquid and gas phases, and a charge balance on the ions in solution. Andrews and Graef[191] utilized the relationships outlined above to develop a model to predict the dynamic response of five of the most commonly used process parameters: volatile acids concentration, pH, alkalinity, gas flow rate, and gas composition. Andrews[35] made several simplifying assumptions for this model that included: using the conversion of acetic acid to methane as the sole rate limiting step; assuming that the inhibition arose only from volatile acids; considering the carbon dioxide-bicarbonate system as the only source of buffering capacity; assuming no transport of carbon dioxide to or from a solid phase; assuming no endogenous respiration, no natural organism decay, and no lag period between reaction steps.

Andrews[35] found that the dynamic model could be used to evaluate the stability of anaerobic digestion processes and the capability of various control systems to maintain stability. The best output variable or combination of variables for monitoring fermentation performance depends in part upon the type of overload failure that is most likely to be experienced. The selection of the most appropriate control action as well depends upon the nature of the process deviation. The most commonly used control actions include: reduction or temporarily halting organic loading, dilution of the digester contents, addition of a base such as lime to raise digester pH, and the addition of sludge from a stable healthy digester. Graef and Andrews[192] also proposed a method of pH control by scrubbing carbon dioxide from the digester gas as it was being recycled through the digester.

Andrews[35] presented a number of simulated digester control actions using a dynamic model. Use of pH as the feedback signal, and base addition as the control mechanism proved to be a useful technique for prevention of failure from organic overloading, but not very useful for control of hydraulic or toxic overloading. Another computer simulation using the rate of methane production as the feedback signal and the addition of active sludge from a healthy digester as the control action appeared to be very successful for prevention of failure due to toxic overloading.

Much more work is required to develop the dynamic model as a comprehensive, accurate tool for anaerobic digester design and control.[35] At present, even the basic microbiological processes occurring during anaerobic digestion are not fully understood. The interdependencies of the numerous bioreactions involved in the conversion of complex organic substrates to methane as well as the effects of inhibitory and toxic substances were reviewed in Section IV. The model proposed by Andrews and Graef[32,34-35,190-191] was applied to the anaerobic digestion of municipal and industrial wastes. It was limited to the prediction of general trends of dynamic behavior in a qualitative rather than a quantitative manner. te Boekhorst et al.[193] pointed to the need for comprehensive analysis of wastewaters intended for anaerobic digestion so that a set of initial and boundary conditions would be available for a kinetic model. There are significant differences between various wastewaters.

Hill and Barth[194] modified the dynamic model that Andrews[34] applied to municipal wastes for application to farm livestock manures. They were able to predict dynamic responses during the digestion of poultry and swine manures to within 10% of the actual field data for the parameters of volatile solids, volatile acids and ammonium concentration. Hill and Barth[194] had included a factor to account for inhibition due to high levels of free ammonia in their version of a dynamic model by adding a term to the growth rate equation used by Andrews[35] as follows:

$$\mu = \frac{\hat{\mu}}{1 + \dfrac{K_s\,K_a}{(S_1)\,(H^+)} + \dfrac{(S_1)\,(H^+)}{K_i\,K_a} + \dfrac{(NH_3)}{K_{i2}}} \tag{54}$$

where: (NH_3) = concentration of the unionized ammonia (mass-vol^{-1}); and K_{i2} = inhibition coefficient for free ammonia (mass-vol^{-1})

Obviously, the development of a comprehensive dynamic model of the anaerobic digestion of farm animal manures would be a breakthrough promoting the more widespread adoption of this valuable pollution control and energy recovery technology in agriculture.

VII. DESIGN OF A RETAINED BIOMASS REACTOR FOR FARM WASTES

A major area of recent advance in anaerobic reactor design has been the development of various systems for biomass retention. A number of these designs were reviewed in Section V. The retention of active biomass separate from the effluent stream has enabled large reductions in HRT while still maintaining long biological solids retention times. This leads to greater process stability and increased efficiencies of waste conversion with substantial reduction in the required reactor volume.

Many of the retained biomass designs that were described in Section V were developed for soluble wastewaters and were not practical for application to wastewaters with relatively high levels of particulate solids such as farm animal wastes. As discussed earlier, the sludge bed reactors experienced buildup of nonbiodegradable solid particles that plugged the sludge bed if they were used to treat highly particulate wastewaters. It has also been recommended that "anaerobic filters" be employed only for wastewaters with less than 1% solid particle content to avoid problems of filter blockage and channeling.[41] The downflow stationary-surface fixed-film design has been successfully operated with 5% TS pig manure,[144] but the surface area for bacterial support in this design is limited if the channels are made large enough to prevent blockage by suspended solids.[41] Several of these retained biomass designs have been adapted for highly particulate wastewaters by employing them as the second stage of a two stage fermentation process.[41] The first stage was a conventional CSTR or batch reactor that liquified the solids and converted the complex organic materials to volatile acids. The output of the first stage with low solid particles and high volatile acids content was then delivered to a retained biomass reactor as the second stage to undergo a rapid conversion to methane.

The fluidized-bed bioreactor design has the capacity to handle relatively high levels of suspended solids. However, this design generally requires high flow rates to maintain fluidization of the particle bed, and recycling of the wastewater is necessary if hydrolysis of the suspended particles becomes a rate-limiting step in the conversion.[167] These factors require increased energy inputs. The liquid recycling and pumping system is also prone to blockage if the wastewater contains a large component of solid particles.

A mechanically agitated, suspended particle, fixed-biomass (SPFB) anaerobic reactor was demonstrated by Blanchard and Gill.[179] This design eliminated some of the problems associated with the treatment of highly particulate wastewaters in retained biomass reactors.

The development of porous biomass support particles[165] that were employed in fluidized-bed reactors opened the way for the use of biomass support particles in other reactor designs such as in mechanically agitated reactors. This type of support particle allows the growth and attachment of biomass within the matrix of the particle voids providing protection from the effects of surface shearing forces from high liquid velocities, interparticle collisions, and particle/impeller collisions. The retention of support particles separate from the reactor effluent can be made more efficient if relatively large support particles are employed. As well, when highly porous support particles ($\geq$ 80% voids) become filled with bacterial growth, they approach the specific gravity of wet biomass (approximately 1.10).[165] The resulting relatively large but low density particles can be suspended in reactor vessels by comparatively low speed, energy efficient impellers. Blanchard and Gill[179] employed 1.5 $\times$ 1.5 $\times$ 0.75 cm rectangular slabs of reticulated nylon fiber mat with a porosity of 96% as biomass support particles. This shape with one smaller dimension was intended to reduce any possible limiting effect on the reaction rate by diffusional limitation of substrate to the underlying layers of bacteria in the support particles. The two larger dimensions facilitated support particle separation from the digester effluent by baffles on the effluent outlet that were not prone to blockage by manure solids.

The laboratory scale reactors used for the manure digestion trials were constructed of cylindrical, flat-bottomed, polyvinyl chloride (PVC) vessels with a 29.5 cm I.D. and 43.1 cm depth. These provided a 25 ℓ digestion capacity when filled to a 38 cm depth allowing for the volume displaced by the impeller and baffles, etc. A variety of agitation devices were tested in different combinations and positions in a test tank to select an energy-efficient particle suspension system with a minimum impeller rotational speed. The final arrangement chosen was a 15.24 cm diameter, six-bladed, disc-turbine impeller located on a centrally mounted, vertical shaft located at one third the liquid depth off the tank bottom operated at 122 rmp. Three 2.5 cm wide vertical wall baffles were spaced 90° apart around the inside wall of the reactor. A 4.5 cm diameter plexiglas loading tube from a horizontally mounted hydraulically driven continuous loading cylinder took the place of a fourth baffle as depicted in Figure 11. The SPFB digesters were equipped with 1.6 ℓ (volume with no interparticle voids) of the porous biomass support particles for the digestion tests. Two other similar digesters were constructed to be operated as continuous stirred-tank (CST) digesters without support particles. These served as controls for comparison. All four digesters were maintained at 35°C throughout the experiment. Initially, fermentations in the four digesters were started under identical conditions and inoculated with the same seed from a municipal wastewater anaerobic digester. The digesters were operated with separated dairy manure slurry at a HRT of 25 days to establish an active population of anaerobes. The four digesters were then carefully monitored over a 6 week period while the HRT was decreased at regular intervals from 25 to 2.5 days. The separated dairy manure contained 5.9% TS, 4.5% volatile solids, and 3.8% suspended solid particles.

Measurements of gas production, gas composition, pH, alkalinity, total solids, volatile solids, and volatile acids were performed daily on each of the four digesters. Figure 12 presents the total gas production rate, percentages of CH_4 and CO_2, alkalinity, pH, and the concentration of total volatile acids. The gas production, gas composition, alkalinity, and pH were similar for all digesters at HRT's from 25 to 10 days. Upon reduction of the HRT to 5 days, there was a divergence in behavior between the CSTR's and the SPFB reactors. The CSTR's, as would be expected, rapidly moved toward failure due to the effects of bacterial washout while the SPFB reactors maintained stable fermentation. The values of the process parameters at the termination of the experiment are presented in Table 5. Figure 12 illustrates a trend toward greater process stability in the SPFB digesters as indicated by the gas production rate, gas composition, alkalinity, and pH. This phenomenon would be predicted by kinetic theory for a reactor with a greater biological solids retention time. The

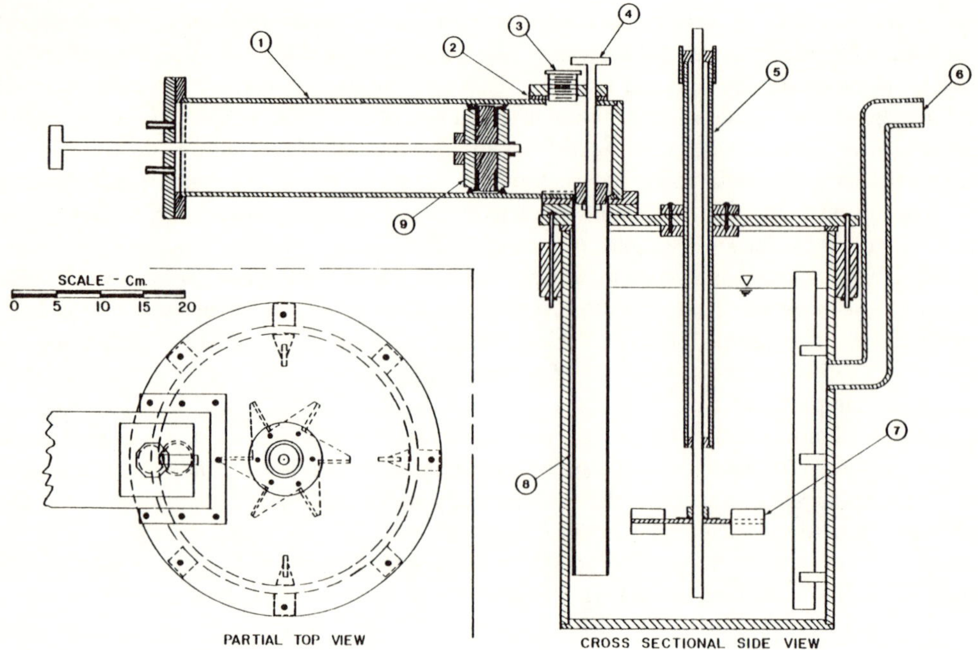

FIGURE 11. Suspended-particle, fixed-biomass digester with continuous loader attached. (1) Loading cylinder. (2) Plexiglas plate. (3) PVC end-plug. (4) Steel shaft with stopper. (5) Impeller shaft casing. (6) Effluent overflow pipe. (7) Disc-bladed turbine impeller. (8) Inlet tube from loading cylinder. (9) Loading piston and shaft.

levels of volatile acids shown in Figure 12 for the CSTR are typical. Initially at HRT's > 5 days, the concentration of volatile acids in the CSTR's was low (<200 mg HAc/ℓ) since nearly all of the volatile acids produced were converted to gas. At HRT's < 5 days, the level of volatile acids rose rapidly to almost 9000 mg as HAc/ℓ indicating washout of the methanogenic bacteria. The levels of volatile acids in the SPFB reactors were significantly higher (up to 1000 mg as HAc/ℓ) at HRT > 5 days. This suggests a greater rate of hydrolysis in the SPFB reactors since the rate of volatile acid conversion to gas was equal in both types of reactor. In spite of the larger capacity for volatile acid production in the SPFB reactors, they did not fail at HRT < 5 days indicating that the methanogenic bacterial population was retained in the porous biomass support particles. The volatile acid levels increased in the SPFB reactors at short HRT's probably because the reactors were not at a steady-state operation due to the brief duration of the experiment. The numbers of acetogenic bacteria would be expected to increase more quickly than the methanogens immediately after increasing the digester loading rate since the growth response of the methanogenic bacteria is the slowest. Over a longer period of operation at the increased loading rate, the numbers of methanogens would be expected to reach a balanced level with the acetogenic bacteria, and the concentration of volatile acids would be reduced.

A series of particle suspension tests were performed in a glass tank equipped with a system of interchangeable mixing impellers, baffles, and other auxiliary mixing devices (Table 6) to select a mixing system that would minimize the particle suspension power inputs and the physical severity of support particle impeller collisions.[195] Mixing power inputs for various arrangements of mixing devices and mixing shaft speed were compared using a dynanometer attached to the mixing shaft. A sketch of the particle suspension test tank without the mixing motor and dynamometer attached is illustrated in Figure 13.

The selection of particle suspension method is critical if reactors are to be expected to

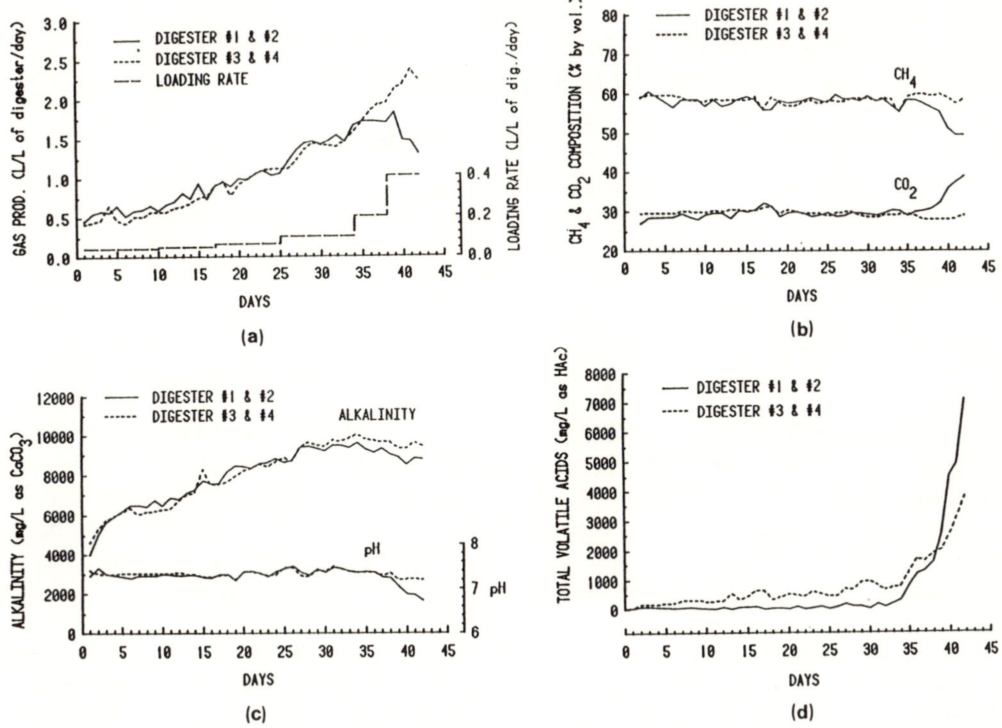

FIGURE 12.   Comparisons of dairy manure fermentations in CSTR's and SPFB reactors. (a) Average daily gas production for the two CSTR's and the two SPFB reactors; (b) average daily gas compositions; (c) average daily pH and alkalinity measurements; and (d) average daily total volatile acids concentrations. Digesters 1 and 2 were CSTR's, and digesters 3 and 4 were SPFB reactors.

Table 5
CST AND SPFB FERMENTATION PARAMETERS AT THE TERMINATION (2.5 DAY HRT) OF THE TRIAL DIGESTION OF DAIRY MANURE

Parameter	CST Reactors (average for digestion 1 and 2)	SPFB Reactors (average for digestion 3 and 4)
Gas production rate (ℓ/ℓ of digester/day)	1.29	2.23
$\%CH_4$	49.1	58.3
$\%CO_2$	38.8	29.0
Alkalinity (mg as $CaCO_3/\ell$)	8746	9399
Total volatile acids (mg as HAc/ℓ)	7041	3702

produce a net positive energy balance. The principles of mechanical agitation as applied to particle suspension have been reviewed by a number of authors.[178,196-198] Disc-bladed turbines were chosen for testing in this application as opposed to paddle impellers or propellers. Paddle impellers are seldom applicable for particle suspension since they push the liquid radically and tangentially ahead of the blades without producing adequate stream velocities for particle suspension.[196] Propellers require higher liquid velocity heads than turbines for a given circulation rate in a mixing vessel. Their circulation capacity drops and power

Table 6
**MIXING DEVICES USED IN VARIOUS COMBINATIONS FOR
PARTICLE SUSPENSION TESTS**

Mixing device	Description
Disc-turbine impellers	10.2 cm diameter; 2.0 cm blade length; 2.5 cm blade height
(6-bladed)	12.7 cm diameter; 2.5 cm blade length; 3.2 cm blade height
	15.2 cm diameter; 3.1 cm blade length; 3.8 cm blade height
	17.8 cm diameter; 3.6 cm blade length; 4.5 cm blade height
Baffles	40 cm height; 2.5 cm width (set of four baffles)
Partial baffles	7.6 cm height; 5 cm width (set of six baffles installed vertically around the circumference of the tank floor)
Draught tubes	Three different diameters with several lengths in each diameter

consumption rises rapidly in high viscosity fluids since the high fluid velocity heads diminish rapidly in such liquids. Since turbines are typically operated at lower speeds (15 to 400 rpm) than propellers (400 to 1750 rpm) and move a greater mass of liquid at lower velocity, they present less risk of attrition of entrapped bacteria from support particle collision with the impeller. The capability of a turbine to maintain constant power consumption independent of viscosity[199] is a desirable feature for this application since the apparent viscosity of wastewaters such as manure slurries vary with composition.

Mixing tests indicated that a turbine impeller located at about one third the digester height from the tank bottom produced the most efficient suspension for the porous support particles.

Mixing tests were also carried out with solid nylon cubes of the same dimensions as the porous particles but with a density approximating the porous particles filled with biomass (1.13 g/cm^3). With these particles, the best energy efficiencies for complete nonhomogeneous particle suspension were observed when the impellers were in the bottom location, and the best efficiencies for complete homogeneous suspension were achieved at any location from bottom to one third off-bottom. For all the test combinations with the solid particles, the middepth impeller location required the highest suspension energy inputs. Power and impeller speeds required to achieve a given level of particle suspension in all cases decreased as the diameter of the impeller was increased. The design chosen for the SPFB reactors employed the 15.24 cm diameter turbine mounted at one third the liquid depth off the bottom driven at a speed of 122 rpm. Table 7 displays the test results that represent the minimum and maximum range of power and speed required for particle suspension with this arrangement.

For a given impeller diameter, equations expressing the square root of dynamometer torque as linear functions of impeller speed were identical regardless of the impeller location or type of particle being suspended. The differences in power requirements at different locations with the same impeller were a consequence of differences in speed required for a given level of particle suspension. Particle suspension performance was independent of apparent viscosity using water, gelatin, and CMC solutions as the suspending fluids. For these liquids, the impeller speeds and mixing power inputs for either complete nonhomogeneous, or, complete homogeneous particle suspension states were the same. The liquid depth to diameter ratio in the particle suspension test tank was 1.2:1. It is anticipated that particle suspension would be more efficient with larger diameter impellers in tanks with depth to diameter ratios less than 1.2:1. Calculations based on the experimental data demonstrated that energy requirements for particle suspension in farm scale reactors would be relatively small compared to the major energy flows (digester heating loads and gas energy production).[195]

The SPFB design exhibited the typical advantages of other retained biomass reactors (stable, high reaction rates) as well as features suitable to the anaerobic digestion of wastewater with appreciably high levels of suspended organic solids. More complete details of this design are presented by Blanchard and Gill.[179,195]

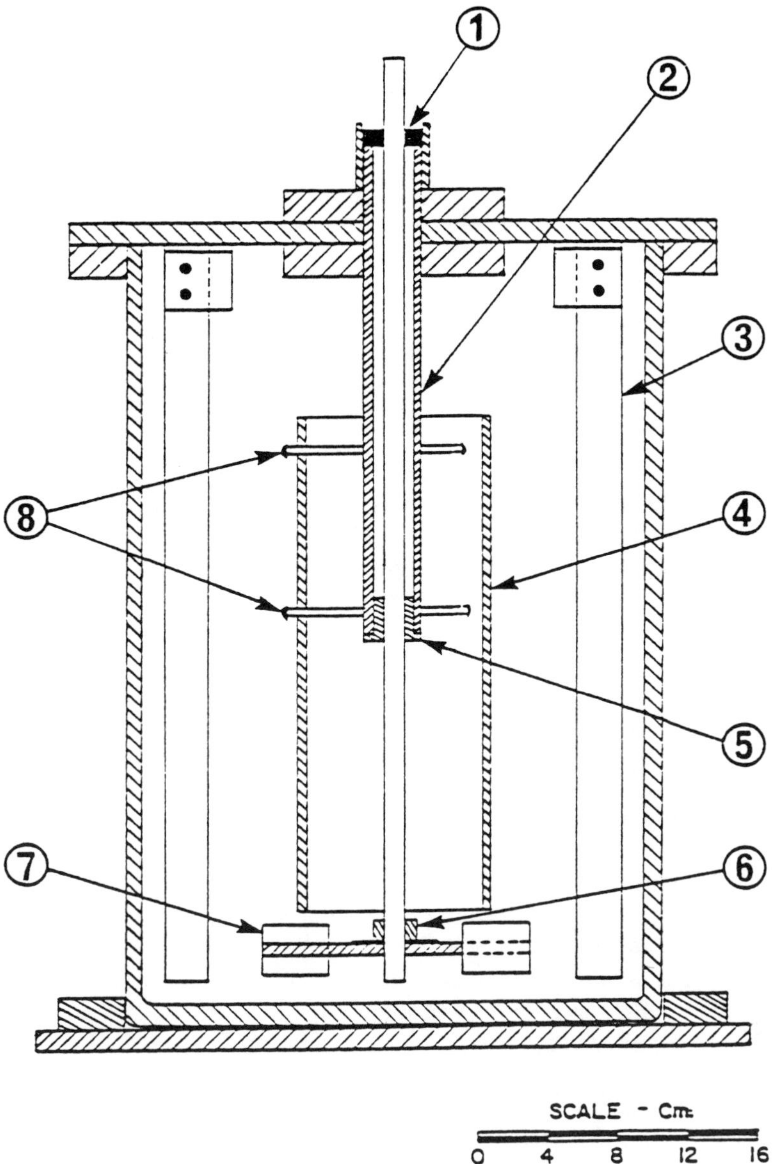

FIGURE 13.   Sketch of the particle suspension test tank with some of the mixing apparatus installed. (1) Tapered roller bearing. (2) Shaft casing. (3) Baffle. (4) Draught tube. (5) Plastic journal bearing. (6) Steel collar and flange. (7) Impeller. (8) Draught tube retaining bolts.

VIII. SUMMARY

Significant advantages can be expected from the application of anaerobic digestion to agricultural wastes. Typically, the pollution potential of livestock or crop wastes can be reduced to less than one tenth of their predigested values. At the same time, a net energy return in the form of methane gas can be achieved. The fertilizer value of the effluent from a farm waste digester is nearly equivalent to the raw waste while malodors and bacteria pathogenic to livestock are greatly reduced. However, other factors have discouraged the

Table 7
RANGE OF POWER INPUTS AND IMPELLER SPEEDS FOR THE 15.2 CM DIAMETER TURBINE AT THE 1/3RD OFF-BOTTOM MOUNT

	Complete suspension		Complete homogeneous suspension	
	Impeller speed (rpm)	Power (W)	Impeller speed (rpm)	Power (W)
Reticulated nylon cuboids (96% voids)	75	0.7	120	2.7
Solid nylon cuboids (Sp. Gr. = 1.13)	160	6.0	190	9.9

widespread application of anaerobic digesters on farms. These include: the high capital cost of construction for the large volumes required for conventional 10 to 30 day HRT reactors; the requirement for skilled operating personnel, and the lack of a comprehensive dynamic model of the anaerobic digestion process. A dynamic model could be used to maintain efficient operation despite variable fermentation conditions brought about by shock loadings and chemical and physical variations in the reactor feed.

The adoption of a comprehensive set of fermentation parameters, including a thorough assessment of the particular reactor feed in question, would help to correlate the results of different studies and would lead to a more rapid expansion of the data base required to develop a dynamic kinetic model. Advances in anaerobic digestion have been a result of either increased stability and rates of the biological reactions involved, or, process intensification by increasing the densities of active bacteria in the reactor. The biological process can be optimized through elucidation of the kinetics of substrate conversion and biological growth and by illuminating the interrelated effects of various growth stimulating, growth inhibiting, and toxic agents that frequently occur in an anaerobic digester. Advances in the understanding of the mechanisms of inhibition by excessive concentrations of such naturally occurring materials as alkali metal cations, heavy metals, volatile acids, ammonia, H_2, and sulfides, etc. have resulted in improved techniques for preventing digestion failures. These advances have also contributed towards the development of a dynamic model of the anaerobic digestion process.

The development of continuous bioreactor designs that retain the active microorganisms separated from the effluent stream has resulted in shortened HRT's and more stable reactors. Reduced HRT's permit smaller reactor volumes which result in reduced construction costs. HRT's have been greatly reduced from the 10 to 30 day HRT's required for conventional CSTR's. Biomass retention systems have been based on the natural ability of the symbiotic population of anaerobic bacteria to form flocs or to attach themselves as films to solid surfaces. Many of these established retained biomass reactor designs (i.e., the anaerobic contact process, sludge blanket reactors, upflow anaerobic filters, and FBFF reactors) have been applied mainly for solubilized wastewaters since they are not as suitable to digestion of high particulate wastewaters such as livestock manures. However, retained biomass designs are being developed for application to high particulate agricultural wastes.

One approach to adopt these designs for farm waste digesters has been to separate the acetogenic phase of the anaerobic digestion from the methanogenic phase so that the rate limiting methane forming stage can be accomplished separately in a retained biomass reactor after the suspended particles have been liquidifed or removed during the first reaction stage in a conventional CSTR. However, the existing SFFR design has shown potential for application to high particulate agricultural wastes as a single stage reactor.

Porous biomass support particles that entrap the active bacteria within their internal void spaces and provide protection from attritional losses due to physical collisions and abrasions are suitable for the development of new retained biomass reactor designs applicable to high particulate wastewaters. One example of this is a SPFB reactor design[179,195] that employed porous nylon fiber cuboids as biomass support particles maintained suspended in a continuous anaerobic digester by a mechanical turbine impeller. Lab-scale trials of this reactor design using a 5.9% total solids separated dairy manure demonstrated an ability to operate efficiently and economically at reduced HRT's.

It should be emphasized that the development of a dynamic kinetic model is a major requirement for making anaerobic digestion widely accessible as a practical on-farm technology. It would ensure stable optimum reactor operation in a farm situation with a minimum of operator expertise. Computer automated process monitoring and control could be realized with the development of a dynamic model.

IX. GLOSSARY OF TERMS

BOD   — biological oxygen demand
BOD_L — ultimate biological oxygen demand
BSRT — biological solids retention time
CST   — continuous stirred-tank
CSTR — continuous stirred-tank reactor
COD — chemical oxygen demand
FBFF — fluidized-bed, fixed-film
HRT   — hydrualic retention time
SFFR — stationary fixed-film reactor
SPFB — suspended-particle, fixed-film
TS   — total solids
UASB — upflow anaerobic sludge blanket
VA   — volatile acids
VS   — volatile solids
VSS   — volatile suspended solids

REFERENCES

1. **Toerien, D. E. and Hattingh, W. H. J.,** Anaerobic digestion. I. The microbiology of anaerobic digestion. *Water Res.* 3, 385, 1965.
2. **Byrant, M. P.,** The microbiology of anaerobic degradation and methanogenesis with special reference to sewage, in *Symp. Microbial Energy Conversion,* Schlegel, H. C. and Barnea, J., Eds., Pergamon Press, Elmsford, N.Y., 1977, 107.
3. **Jeris, J. S. and McCarty, P. L.,** The biochemistry of methane fermentation using C14 tracers, *J. Water Pollut. Control Fed.,* 37(2), 178, 1965.
4. **Eckenfelder, W. W., Jr.,** Mechanisms of sludge digestion, *Water Sewage Works,* 114(6), 207, 1967.
5. **Pos, J., te Boekhorst, R. H., and Pavlicik, V.,** The design and operation of a 1400L anaerobic digester for hog manure, Paper No. 71-17, Canadian Society of Agricultural Engineers, 1979, 1.
6. **Hobson, P. N. and Robertson, A. M.,** *Waste Treatment in Agriculture,* Applied Science, London, 1977, chap. 8.
7. **Jewell, W. J., Davis, H. R., Gunkel, W. W., Lathwell, D. J., Martin, J. H., Jr., McCarty, T. R., Morris, G. R., Price, D. R., and Williams, D. W.,** Bioconversion of Agricultural Wastes for Pollution Control and Energy Conservation, TID-27164, U.S. Department of Commerce, National Technical Information Service, Washington, D.C., 1976.

8. **McCarty, P. L.,** Anaerobic waste treatment fundamentals. I. Chemistry and microbiology, *Public Works,* 95(8), 107, 1964.
9. **Cowley, I. D. and Wase, D. A. J.,** Anaerobic digestion of farm wastes: a review, I, *Process Biochem.,* 16(5), 28, 1981.
10. **Iannotti, E. L., Porter, J. H., Fischer, J. R., and Sievers, D. M.,** Changes in swine manure during anaerobic digestion, *Dev. Ind. Microbiol.,* 19, 519, 1979.
11. Midwest Plan Service, *Livestock Waste Management with Pollution Control,* MWPS-19, Iowa State University, Ames, Iowa, 1975, 46.
12. Midwest Plan Service, *Livestock Waste Facilities Handbook,* MWPS-18, Iowa State University, Ames, Iowa, 1975, 5.
13. **Sobel, A. T. and Muck, R. E.,** Energy in animal manures, ASAE Paper No. NA81-221, American Society of Agricultural Engineers, St. Joseph, Michigan, 1981, 1.
14. **McCarty, P. L.,** Anaerobic waste treatment fundamentals. II. Environmental requirements and control, *Public Works,* 95(10), 123, 1964.
15. **Hawkes, D. L. and Horton, R.,** Anaerobic digester design — fundamentals, III, *Process Biochem.,* 16(2), 10, 1981.
16. **Lawrence, A. W.,** Application of process kinetics to design of anaerobic processes, in *Anaerobic Biological Treatment Processes,* Gould, R. F., Ed., American Chemical Society, Washington, D.C., 1971, 163.
17. **Persson, S. P. E. and Bartlett, H. D.,** Rigid wall mixed reactors for dairies, in *Proc. Methane Technol. Agric. Conf.,* NRAES Cornell University, Ithaca, New York, 1981, 1.
18. **Cowley, I. D. and Wase, D. A. J.,** Anaerobic digestion of farm wastes; a review, II, *Process Biochem.,* 16(6), 16, 1981.
19. **Ross, W. R. and Smollen, M.,** Engineering problems with the anaerobic digestion of soluble organic wastes, in *Anaerobic Digestion — 1981,* Huges, D. E., et al., Eds., Elsevier Biomedical Press, Amsterdam, 1981, 89.
20. **Anonymous,** Anaerobic sludge digestion, *J. Water Pollut. Control Fed.,* 38(10), 1683, 1966.
21. **Malina, J. E. and Miholitis, E. M.,** New developments in anaerobic digestion of sludges, in *Advances in Water Quality Improvement,* Gloyna, E. F. and Eckenfelder, W. W., Jr., Eds., University of Texas Press, Austin, Texas, 1968, 355.
22. **Zoltak, J. and Gram, A. L.,** High rate digester mixing study using radiosotope tracers, *J. Water Pollut. Control Fed.,* 47, 79, 1975.
23. U.S. Environmental Protection Agency, Process Design Manual for Sludge Treatment and Disposal, EPA 625/1-79-011, Washington, D.C., September, 1979.
24. **Hashimoto, A. G., Chen, Y. R., Varel, V. H., and Prior, R. L.,** Anaerobic fermentation of animal manures, ASAE Paper No. 79-4066, American Society of Agricultural Engineers, St. Joseph, Michigan, 1979, 1.
25. **Haliburton, J. D.,** The annual energy balance for anaerobically digested hog manure, unpublished paper, Agricultural Engineering Department, Univ. of Manitoba, Winnipeg, Canada.
26. **Stafford, D. A. and Etheridge, S. P.,** Farm wastes, energy production and the economics of farm anaerobic digesters, in *Anaerobic Digestion — 1981,* Hughes, D. E., et al., Eds., Elsevier Biomedical Press, Amsterdam, 1982, 255.
27. **Hayes, T. D., Jewell, W. J., Chandler, J. A., Dell'Orto, S., Fanfoni, K. J., Leuschner, A. P., and Sherman, D. F.,** Methane generation from small scale farms, in *Proc. of 1979 Semin. Biogas Alcohol Fuels Prod.,* JG Press, Emmaus, Pa., 1980, 88.
28. **Horton, H. R.,** The implication of engineering design on anaerobic digester systems, in *Anaerobic Digestion,* Stafford, D. A., Wheatley, B. I., and Hughes, D. E., Eds., Applied Science, London, 1980, chap. 17.
29. **Stafford, D. A., Hawkes, D. L., and Horton, H. R.,** *Methane Production from Waste Organic Matter,* CRC Press, Boca Raton, Fla., 1980, chap. 2.
30. **Colleran, E., Barry, M., Wilkie, A., and Newell, P. J.,** Anaerobic digestion of agricultural wastes using the upflow anaerobic filter design, *Process Biochem.,* 17(2), 12, 1982.
31. **McCarty, P. L.,** Aanerobic treatment of soluble wastes, in *Advances in Water Quality Improvement,* Gloyna, E. F. and Eckenfelder, W. W., Jr., Eds., University of Texas Press, Austin, Tex., 1968, 336.
32. **Andrews, J. F.,** Control strategies for the anaerobic digestion process. I and II, *Water Sewage Works J.,* 122(3,4), 62, 74, 1975.
33. **Buvet, R.,** Biomeh tanation: units of measurements, in *Anaerobic Digestion — 1981,* Hughes, D. E., et al., Eds., Elsevier Biomedical Press, Amsterdam, 1982, 429.
34. **Andrews, J. F.,** Dynamic model of the anaerobic digestion process, *J. Sanit. Eng. Div., Am. Soc. Civ. Eng.,* 95(SA1), 95, 1969.
35. **Andrews, J. F.,** Dynamic models and control strategies for biological wastewater treatment processes, in *Mathematical Modeling for Water Pollution Control Processes,* Keinath, T. M. and Wanielista, M. P., Eds., Ann Arbor Science Publishers Inc., Ann Arbor, Mich., 1975, 319.

36. **Bungay, H. R.,** Dynamic analysis of microbial systems, *Water Sewage Works J.,* 114(5), 190, 1967.
37. **van den Heuvel, J. C. and Zoetemeyer, R. J.,** Stability of the methane reactor: a simple model including substrate inhibition and cell recycle, *Process Biochem.,* 17(3), 14, 1982.
38. **Herbert, D.,** A theoretical analysis of continuous culture systems, in *Continuous Culture of Microorganisms,* Society of Chemical Industries (S.C.J.) Monograph No. 12, 1961, 21.
39. **Pretorius, W. A.,** Anaerobic digestion. III. Kinetics of anaerobic fermentations, *Water Res.,* 3, 545, 1969.
40. **Demuynck, M., Perez-Aguilar, I., Naveau, H., and Nyns, E-J.,** Biomethanation: units and measurements, in *Anaerobic Digestion — 1981,* Hughes, D. E. et al., Eds., Elsevier Biomedical Press, Amsterdam, 1982, 107.
41. **Callander, I. J. and Barford, J. P.,** Recent advances in anaerobic digestion technology, *Process Biochem.,* 18(4), 24, 1983.
42. **Woods, J. L. and O'Callaghan, J. R.,** Mathematical modelling of animal waste treatment, *J. Agric. Eng. Res.,* 19, 245, 1974.
43. **Lawrence, A. W. and McCarty, P. L.,** Kinetics of methane fermentation in anaerobic treatment. II, *J. Water Pollut. Control Fed.,* 41(2), R1, 1969.
44. **Lawrence, A. W. and McCarty, P. L.,** Unified basis for biological treatment design and operation, *J. Sanit. Eng. Div. Am. Soc. Civ. Eng.,* 91(SA3), 757, 1970.
45. **Atkinson, B. and Daoud, I. S.,** Diffusion effects within microbial films, *Trans. Inst. Chem. Eng.,* 48, T245, 1970.
46. **Kotze, J. P., Thiel, P. G., and Hattingh, W. H. J.,** Anaerobic digestion. II. The characterization and control of anaerobic digestion, *Water Res.,* 3, 459, 1969.
47. **Kroeker, E. J., Schulte, D. D., Sparling, A. B., and Lapp, H. M.,** Anaerobic treatment process stability, *Water Pollut. Control Fed.,* 51, 718, 1979.
48. **Morris, G. R., Jewell, W. J., and Loehr, R. C.,** Anaerobic fermentation of animal wastes: a kinetic design evaluation, in *Proc. 32nd Ind. Waste Conf.* Ann Arbor Science, Ann Arbor, Michigan, 1978, 689.
49. ASAE, *Agricultural Engineers Yearbook,* ASAE, St. Joseph, Mich., 1983, 436.
50. **O'Callaghan, J. R., Dodd, V. A., and Pollock, K. A.,** The long term management of animal manures, *J. Agric. Eng. Res.,* 18, 1, 1973.
51. **Ghose, T. K. and Das, D.,** Maximization of energy recovery in the biomethanation process. II. Use of a mixed residue in batch systems, *Process Biochem.,* 17(1), 39, 1982.
52. **Hills, D. J.,** Biogas from dairy and carbonaceous wastes at high solids, ASAE, Paper No. 79-4582, American Society of Agricultural Engineers, St. Joseph, Michigan, 1979, 1.
53. **Hobson, P. N., Bousfiels, S., and Summers, R.,** Anaerobic digestion of organic matter, *Crit. Rev. Environ. Control,* 4(2), 131, 1974.
54. **Zeikus, J. G.,** Fate of lignin and related aromatic substrates in anaerobic environments, in *Lignin Biodegradation Microbiology, Chemistry and Potential Applications,* Vol. 1., Kirk, T. K., Higuchi, T., and How-min Chang, Eds., CRC Press, Boca Raton, Fla., 1980, 101.
55. **Hungate, R. E.,** *The Rumen and Its Microbes,* Academic Press, New York, 1966, chap. 10.
56. **Gossett, J. M. and McCarty, P. L.,** Heat treatment of refuse for increasing anaerobic biodegradability, in *Biochemical Engineering — Energy, Renewable Resources, and New Foods,* AIChE Symp. Series, No. 158, Vol. 72, American Institute of Chemical Engineers, New York, 1976, 64.
57. **McCarty, P. L., Young, L. Y., Gossett, J. M., Stuckey, D. C., and Healy, J. B., Jr.,** Heat treatment for increasing methane yields from organic materials, in *Microbial Energy Conversion — A Seminar Proceedings,* Pergamon Press, Elmsford, N.Y., 1977, 179.
58. **Millet, M. A., Baker, A. J., and Satter, L. D.,** Pretreatments to enhance chemical, enzymatic, and microbial attack of cellulosic materials, in *Biotechnology and Bioeng. Symp. No. 5,* John Wiley & Sons, New York, 1975, 193.
59. **Chynoweth, D. P., Ghosh, S., and Packer, M. L.,** Anaerobic processes, *J. Water Pollut. Control Fed.,* 51(6), 1200, 1979.
60. **Klein, S. A.,** Anaerobic digestion of solid wastes, *Compost Sci.,* 13(1), 6, 1972.
61. **Harper, J. P., Ngoddy, P. O., and Garrish, J. B.,** Enhanced treatment of livestock wastewater. II. Enhancement of treatment by solids removal, *J. Agric. Eng. Res.,* 19, 353, 1974.
62. **Frecks, G. A. and Gilbertson, C. B.,** The effect of ration on engineering properties of beef cattle manure, ASAE 73-422, American Society of Agricultural Engineers, St. Joseph, Michigan, 1973, 1.
63. **Loehr, R. C.,** *Agricultural Waste Management,* Academic Press, New York, 1974, chap. 9.
64. **Verstraete, W., Baere, L., and Rozzi, A.,** Phase separation in anaerobic digestion: motives and methods, *Trib. Cabadeau,* 34, 367, 1981.
65. **Jewell, W. J., Dell'Orto, S., Fanfoni, K. J., Fast, S. J., Gotting, E. J., Jackson, D. A., and Kabrick, R. M.,** Agricultural and high strength wastes, in *Anaerobic Digestion — 1981,* Hughes, D. E., et al., Eds., Elsevier Biomedical Press, Amsterdam, 1982, 151.
66. **McCarty, P. L., Jeris, J. S., and Murdock, W.,** Individual volatile acids in anaerobic treatment, *J. Water Pollut. Control Fed.,* 32(12), 1501, 1963.

67. **Massey, M. L. and Pohland, F. G.,** Phase separation of anaerobic stabilization by kinetic controls, *J. Water Pollut. Control Fed.,* 50(9), 2204, 1978.
68. **Borchardt, J. A.,** Anaerobic phase separation by dialysis technique, in *Anaerobic Biological Treatment Processes — Advances in Chemistry Series 105,* Gould, R. F., Ed., American Chemical Society, Washington, D.C., 1971, 108.
69. **Pohland, F. G. and Mancy, J. H.,** The use of pH and pE measurements during methane biosynthesis, *Biotechnol. Bioeng.,* 11(4), 683, 1969.
70. **Schaumberg, F. F. and Kirsch, E. J.,** Anaerobic simulated mixed culture system, *Appl. Microbiol.,* 14, 761, 1966.
71. **Ghose, T. K. and Bhadra, A.,** Maximization of energy recovery in biomethanation process. I. Use of cow dung as substrate in multireactor systems, *Process Biochem.,* 16(6), 23, 1981.
72. **Ghosh, S., Conrad, J. R., and Klass, D. L.,** Anaerobic acidogenesis of wastewater sludge, *J. Water Pollut. Control Fed.,* 47(1), 30, 1975.
73. **Pohland, F. G. and Ghosh, S.,** Developments in anaerobic stabilization of organic wastes — the two-phase concept, *Environ. Lett.,* 1, 4, 1971.
74. **Smith, R. E., Reed, M. J., and Kiker, J. T.,** Two-phase anaerobic digestion of swine waste, *Trans. ASAE,* 20(6), 1123, 1977.
75. **Norrman, J. and Forstell, B.,** Anaerobic wastewater treatment in a two-stage reactor of a new design, in *Proc. 32nd Ind. Waste Conf.,* Ann Arbor, Mich., 1978, 387.
76. **McCarty, P. L.,** Anaerobic waste treatment fundamentals. III. Toxic materials and their control, *Public Works,* 95(1), 91, 1964.
77. **Kugelman, I. J. and Chin, K. K.,** Toxicity, synergism, and antagonism in anaerobic waste treatment processes, in *Anaerobic Biological Treatment Processes, Advances in Chemistry Series 105,* Gould, R. F., Ed., American Chemical Society, Washington, D.C., 1971, 53.
78. **Kugelman, I. J. and McCarty, P. L.,** Cation toxicity and stimulation in anaerobic waste treatment, *J. Water Pollut. Control Fed.,* 37(1), 97, 1965.
79. **Falk, I. S.,** The role of certain ions in bacterial physiology. A review (studies on salt action), *Abstr. Bacteriol.,* 7, (33, 87, 133), 1923.
80. **Buswell, A. M., Boruff, C. S., and Wiesman, C. K.,** Anaerobic stabilization of milk wastes, *Ind. Eng. Chem.,* 24, 1423, 1949.
81. **Fair, G. M. and Carlson, C. L.,** Sludge digestion — reaction and control, *J. Boston Soc. Civ. Eng.,* 14, 82, 1927.
82. **Shuyler, L. R., Clark, D. A., Barth, J., and Smith, D. D.,** Excretion of salts by feedlot cattle in response to variations in concentrations of sodium chloride added to their ration, in *Managing Livestock Wastes, Proc. 3rd Int. Symp. Livestock Wastes,* ASAE, St. Joseph, Michigan, 1975, 336.
83. **Safley, L. M., Barker, J. C., and Westerman, P. W.,** Characteristics of fresh dairy manure, *Trans. ASAE,* 27(4), 1150, 1984.
84. **Anderson, G. K., Donnelly, T., and McKeown, K. J.,** Identification and control of inhibiton in the anaerobic treatment of industrial wastewaters, *Process Biochem.,* 17(4), 28, 1982.
85. **Mosey, F. E. and Hughes, D. A.,** Toxicity of heavy metal ions to anaerobic digestion, *J. Water Pollut. Control Fed.,* 74, 18, 1975.
86. **Lawrence, A. W. and McCarty, P. L.,** The role of sulfide in preventing heavy metal toxicity in anaerobic treatment, *J. Water Pollut. Control Fed.,* 37, 392, 1965.
87. **Baker, H. A.,** *Bacterial Fermentations,* Chapman and Hall, Ltd., New York, 1956, chap. 1.
88. **McCarty, P. L., Jeris, J. S., and Murdock, W.,** Individual volatile acids in anaerobic treatment, *J. Water Pollut. Control Fed.,* 35(12), 1501, 1963.
89. **Duarte, A. C.,** Inhibition Modelling in Anaerobic Digesters, Ph. D. thesis, University of Newcastle-upon-Tyne, Newcastle, England.
90. **Keefer, C. E., and Urtes, H. C.,** Digestion of volatile acids, *J. Water Pollut. Control Fed.,* 35, 334, 1963.
91. **Clark, R. H. and Speece, R. E.,** The pH tolerance of anaerobic digestion, in *Advances in Water Pollution Research, Proc. 5th Int. Conf., 1970,* Vol. 1., Jenkins, S. H., Ed., Pergamon Press, Elmsford, N.Y., 1971, Section II-27, 1.
92. **Stanier, R. Y., Doudoroff, M., and Adelberg, E. A.,** *General Microbiology,* 2nd ed., Macmillan, New York, 1964, chap. 10.
93. **McCarty, P. L. and McKinney, R. E.,** Salt toxicity in anaerobic digestion, *J. Water Pollut. Control Fed.,* 33(4), 399, 1961.
94. **Sathananthan, S.,** Ammonia Toxicity in Anaerobic Digesters, Masters thesis, University of Newcastle-upon-Tyne, Newcastle, England, 1981.
95. **Sievers, D. M. and Brune, D. E.,** Carbon/nitrogen ratio and anaerobic digestion of swine waste, *Trans. ASAE,* 21, 537, 1978.

96. American Public Health Association, *Standard Methods for the Examination of Water and Wastewater*, 14th ed., Rand, M. C., et al., Eds., American Public Health Association, Washington, D.C., 1976, 278.

97. **Garber, W. F.**, Plant-scale studies of thermophilic digesters at Los Angeles, *Sewage Ind. Wastes*, 26, 1202, 1954.

98. **Kaspar, K. E. and Wuhrman, K.**, Kinetic parameters and relative turnovers of some important catabolic reactions in digester sludge, *Appl. Environ. Microbiol.*, 36(1), 1, 1978.

99. **Shea, T. G., Pretorius, W. A., Coles, R. D., and Pearson, E.**, Kinetics of hydrogen assimilation in the methane fermentation, *Water Res.*, 2, 833, 1968.

100. **Rudolfs, W. and Amberg, H.**, White water treatment. II. Effects of sulfides on digestion, *Sewage Ind. Wastes*, 24, 1278, 1952.

101. **Byrant, M. P., Campbell, L. L., Reddy, C. A., and Crabill, M. R.**, Growth of desulfovibrio in lactate or ethanol media low in sulfate in association with H_2 — utilizing methanogenic bacteria, *Appl. Environ. Microbiol.*, 33(5), 1162, 1977.

102. **Winfrey, M. R. and Zeikus, J. G.**, Effect of sulfate on carbon and electron flow during microbial methanogenesis in freshwater sediments, *Appl. Environ. Microbiol.*, 33(2), 275, 1977.

103. **Decker, K., Jungermann, K., and Thauer, R. K.**, Energy production in anaerobic organisms, *Angew. Chem. Int. Ed. Eng.*, 4(138), 1970.

104. **McCarty, P. L., Kugelman, I. J., and Lawrence, A. W.**, Ion effects in anaerobic digestion, Tech. Rept. No. 33, Department of Civil Engineering, Stanford University, Stanford, Calif., 1964.

105. **Simpson, J. R.**, Some aspects of the biochemistry of anaerobic digestion, in *Waste Treat.*, Issac, P. C. G., Ed., Pergamon Press, Oxford, 1960, 31.

106. **Buswell, A. M., and Pagano, J. F.**, Reduction and oxidation of nitrogen compounds in polluted streams, *Sewage Ind. Wastes*, 24, 897, 1952.

107. **Atkinson, B. and Swilley, E. L.**, The effects of ferric chloride on the removal efficiency of a biological film reactor, *Water Res.*, 1, 687, 1967.

108. **Dague, R. R.**, Application of digestion theory to digestion control, *J. Water Pollut. Control Fed.*, 40(12), 2021, 1968.

109. **Pavoni, J. L., Tenney, W. M., and Echelberger, W. F., Jr.**, Bacterial exocellular polymers and biological flocculation, *J. Water Pollut. Control Fed.*, 44, 414, 1972.

110. **Tenney, M. W. and Verhoff, F. H.**, Chemical and autoflocculation of microorganisms in biological waste water treatment, *Biotechnol. Bioeng.*, 15, 1045, 1973.

111. **Friedman, B. A., Dugan, P. R., Pfister, R. M., and Remsen, C. C.**, Structure of exocellular polymers and their relationships to bacterial flocculation, *J. Bact.*, 98, 1328, 1969.

112. **Jones, C. H., Roth, I. L., and Sanders, W. M.**, Electron microscope study of a slime layer, *J. Bacteriol.*, 99, 316, 1969.

113. **Wilkinson, J. F.**, The extracellular polysaccharides of bacteria, *Bacteriol. Rev.*, 22, 46, 1958.

114. **Characklis, W. G.**, Attached microbial growths. I. Attachment and growth. *Water Res.*, 7, 1113, 1973.

115. **Reid, G. W. and Assenzo, J. R.**, Biological slimes, *Air Water Pollut.*, 5(2—4), 347, 1963.

116. **Atkinson, B.**, Immobilized biomass — a basis for process development in wastewater treatment, in *Biological Fluidised Bed Treatment of Water and Wastewater*, Cooper, P. F. and Atkinson, B., Eds., Ellis Horwood Limited, Chichester, England, 1981, 24.

117. **Kolot, F. B.**, Microbial carriers — strategy for selection, *Process Biochem.*, 16(5), 2, 1981.

118. **Kolot, F. B.**, Microbial carriers — strategy for selection. II, *Process Biochem.*, 16(6), 30, 1981.

119. **Marcipar, A., Cochet, N., Brackenridge, L., and Lebeault, J. M.**, Immobilization of yeasts on ceramic supports, *Biotechnol. Lett.*, 1(2), 65, 1980.

120. **Kornegay, B. H. and Andrews, J. F.**, Kinetics of fixed-film biological reactors, *J. Water Pollut. Control Fed.*, 40(11), R460, 1968.

121. **Heukelekian, H.**, Slime formation in polluted waters. II. Factors affecting slime growth, *Sewage Ind. Wastes*, 28, 78, 1956.

122. **Hartmann, L.**, Influence of turbulence on the activity of bacterial slimes, *J. Water Pollut. Control Fed.*, 39, 958, 1967.

123. **de Vocht, M., van Meenan, P., van Assche, P., and Verstraete, W.**, Influence of sedimentation and adhesion on selection of methanogenic associations, *Process Biochem.*, 18(6), 31, 1983.

124. **Schroepfer, G. J., Fullen, W. J., Johnson, A. S., Ziemke, N. R., and Anderson, J. J.**, Anaerobic contact process as applied to packing house wastes, *Sewage Ind. Wastes*, 27, 460, 1955.

125. **Steffen, A. J.**, Treatment of packing house wastes by anaerobic digestion, in *Advances in Biological Waste Treatment. Proceedings of the Third Conference on Biological Waste Treatment Sponsored by Manhattan College*, Eckenfelder, W. W., Jr. and McCabe, J., Eds., MacMillan, New York, 1963, 126.

126. **Cillie, G. G., Henzen, M. R., Stander, G. J., and Baillie, R. D.**, Anaerobic digestion. IV. The application of the process in waste purification, *Water Res.*, 3, 623, 1969.

127. **van den Berg, L. and Lentz, C. P.,** Food processing waste treatment by anaerobic digestion, in *Proc. 32nd Purdue Industrial Waste Conference,* Ann Arbor Science Publishers Inc., Ann Arbor, Mich., 1977, 252.

128. **Simpson, D. E.,** Investigations on a pilot-plant contact digester for the treatment of a dilute urban waste, *Water Res.,* 5, 523, 1971.

129. **Young, J. C. and McCarty, P. L.,** Technical report No. 87., Grant WP100584, Engineering Research Institute, Iowa State University, Ames, Iowa, 1968.

130. **Stander, G. T.,** Treatment of wine distillery wastes by anaerobic digestion, Purdue Univ. Eng. Bull., Ext. Serv., No. 129(2), Lafayette, Ind., 1967, 892.

131. **Lettinga, G., van Velsen, A. F. M., Hobma, S. W., deZeeuw, W., and Klapwijk, A.,** Use of the upflow sludge blanket (USB) reactor concept for biological wastewater treatment, especially for anaerobic treatment, *Biotechnol. Bioeng.,* 22, 699, 1980.

132. **Pette, K. C. and Versprille, A. I.,** Application of the U.A.S.B. concept for wastewater treatment, in *Anaerobic Digestion — 1981,* Hughes, D. E., et al., Eds., Elsevier Biomedical Press, Amsterdam, 1982, 121.

133. **van den Berg, L., Hamoda, M. F., and Kennedy, K. J.,** Performance of upflow anaerobic sludge bed reactors, in Proc. 3rd Bioenergy R & D Seminar, 1981, Div. Biol. Sci., Natl. Res. Counc., Ottawa, Canada, 1981, 109.

134. **Hall, E. R., Jovanovic, M., and Pejic, M.,** Pilot studies of methane production in fixed-film and sludge blanket anaerobic reactors, in Proc. 4th Bioenery R & D Seminar, Winnipeg, Canada, 1982, 475.

135. **Heertjes, P. M. and van der Meer, R. R.,** Comparison of different methods for anaerobic treatment of dilute wastewaters, *Biotechnol. Bioeng.,* 20, 1577, 1978.

136. **Young, J. C. and McCarty, P. L.,** The anaerobic filter for waste treatment, *J. Water Pollut. Control Fed.,* 41(5), Part 2, R160, 1969.

137. **Colleran, E., Booth, A., Barry, M., Wilkie, A., Newell, P. J., and Dunican, L. K.,** *Bioconversion of Agricultural Wastes into Fuel Gas and Animal Feed,* Palz, W., Chartier, P., and Hall, D. O., Eds., Applied Science, London, 1981, 416.

138. **Young, J. C.,** Performance of anaerobic filters under transient loading and operation conditions, in Anaerobic Filters: An Energy Plus for Wastewater Treatment, Rep. No., ANL/CNSV-TM-50, Argonne Natl. Lab., Argonne, Ill., 1981, 159.

139. **Kennedy, K. J. and van den Berg, L.,** Methane production from piggery-waste using downflow stationay fixed film reactors, in Proc. 4th Bioenergy R & D Seminar, Winnipeg, Canada, 1982, 451.

140. **van den Berg, L. and Kennedy, K. J.,** Factors affecting methane production rates of downflow stationary fixed film reactors, in Proc. 4th Bioenergy R & D Seminar, Winnipeg, Canada, 1982, 457.

141. **van den Berg, L. and Kennedy, K. J.,** Comparison of intermittent and continuous loading of stationary fixed-film reactors for methane production from wastes, *J. Chem. Tech. Biotechnol.,* 32, 427, 1982.

142. **van den Berg, L. and Lentz, C. P.,** Comparison between up- and downflow anaerobic fixed film reactors of varying surface-to-volume ratios for the treatment of bean blanching waste, Paper presented at the 34th Purdue Industrial Waste Conference, 1980, 319.

143. **van den Berg, L. and Lentz, C. P.,** Effect of digester configuration, waste composition and inoculum on rates of production of methane from wastes, presented at 2nd Bioenergy R & D Seminar, March 25, 1980, Ottawa, Canada, 1980.

144. **van den Berg, L., Kennedy, K. J., and Hamoda, M. F.,** Effect of type of waste on performance of anaerobic fixed-film and upflow sludge bed reactors, in Proc. 36th Ind. Waste Conf., Purdue, 1982, 686.

145. **Weber, W. J., Jr.,** Physicochemical treatment of wastewater, *J. Water Pollut. Control Fed.,* 42(1), 83, 1970.

146. **Atkinson, B. and Davies, I. J.,** The completely mixed microbial film fermentor. A method of overcoming wash-out in continuous fermentation, *Trans. Inst. Chem. Eng.,* 50, 208, 1972.

147. **Jeris, J. S., Beer, C., and Mueller, J. A.,** High rate biological denitrification using a granular fluidised-bed, *J. Water Pollut. Control Fed.,* 46(9), 2118, 1974.

148. **Ault, R. G., Hampton, A. N., Newton, R., and Roberts, R. H.,** Biological and biochemical aspects of tower fermentation, *J. Int. Brew.,* 75(3), 260, 1969.

149. **Atkinson, B., Black, G. M., and Pinches, A.,** Process intensification using cell support system, *Process Biochem.,* 15(4), 24, 1980.

150. **Richardson, J. F.,** Incipient fluidization and particulate systems, in *Fluidization,* Davidson, J. F. and Harrison, D., Eds., Academic Press, New York, 1971, 26.

151. **Richardson, J. F. and Zaki, W. N.,** Sedimentation and fluidisation. I, *Trans. Inst. Chem. Eng.,* 32, 35, 1954.

152. **Steinour, H. H.,** Rate of sedimentation, *Ind. Eng. Chem.,* 36, 618, 1944.

153. **Lewis, W. K., Gilliland, E. R., and Bauer, W. C.,** Characteristics of fluidised particles, *Ind. Eng. Chem.,* 41(6), 1104, 1948.

154. **Lewis, W. K. and Bowerman, E. W.,** Fluidization of solid particles in liquids, *Chem. Eng. Prog.,* 48(12), 603, 1952.
155. **Barnea, E. and Mednick, R. L.,** Correlation for minimum fluidization velocity, *Trans. Inst. Chem. Eng.,* 53, 278, 1975.
156. **Ergun, S.,** Fluid flow through packed columns, *Chem. Eng. Prog.,* 48(2), 89, 1952.
157. **Richardson, J. F. and Meikle, R. A.,** Sedimentation and fluidisation. III. The sedimentation of uniform fine particles and of two-component mixtures of solids, *Trans. Inst. Chem. Eng.,* 39, 348, 1961.
158. **Wallis, G. B.,** A simple correlation for fluidisation and sedimentation, *Trans. Inst. Chem. Eng.,* 55, 74, 1977.
159. **Carman, P. C.,** Fluid flow through granular beds, *Trans. Inst. Chem. Eng.,* 15, 150, 1937.
160. **Wilhelm, R. H. and Kwauk, M.,** Fluidization of solid particles, *Chem. Eng. Prog.,* 44, 210, 1948.
161. **Othmer, D. F. and Zenz, F. A.,** *Fluidisation and Fluid-Particle Systems,* Reinhold, New York, 1960, chap. 9.
162. **Sutton, P. M., Shieh, W. K., Kos, P., and Dunning, P. R.,** Dorr-Oliver's Oxitron system fluidized-bed water and wastewater treatment process, in *Biological Fluidised Bed Treatment of Water and Wastewater,* Ellis Horwood Limited, Chichester, England, 1981, 285.
163. **Jewell, W. J. and Cummings, R. J.,** An optimized biological waste treatment process of oxygen utilization, paper presented at the 47th Annual Conference of the Water Pollution Control Federation, Denver, Colorado, October, 1974.
164. **Jewell, W. J., Switzenbaum, M. S., and Morris, J. W.,** Sewage treatment with the anaerobic attached microbial film expanded bed process, paper presented at the 52nd Water Pollution Control Federation Conferene, Huston, Texas, October, 1979.
165. **Atkinson, B., Black, G. M., and Pinches, A.,** The characteristics of solid supports and biomass support particles when used in fluidized beds, in *Biological Fluidized Bed Treatment of Water and Wastewater,* Cooper, P. F. and Atkinson, B., Eds., Ellis Horwood Limited, Chichester, England, 1981, 75.
166. **Atkinson, B.,** Immobilized biomass — a basis for process development, in *Biological Fluidized Bed Treatment of Water and Wastewater,* Cooper, P. F. and Atkinson, B., Eds., Ellis Horwood Limited, Chichester, England, 1981, 22.
167. **Jewell, W. J.,** Development of attached microbial film expanded-bed process for aerobic and anaerobic waste treatement, *Biological Fluidized Bed Treatment of Water and Wastewater,* Cooper, P. F. and Atkinson, B., Eds., Ellis Horwood Limited, Chichester, England, 1981, 251.
168. **Hickey, R. F. and Owens, R. W.,** Methane generation from high-strength industrial wastes with the anaerobic biological fluidized bed, *Biotechnol. Bioeng. Symp.,* 11, 399, 1981.
169. **Switzenbaum, M. S.,** *The AAFEB Reactor for the Treatment of Dilute Organic Wastes,* Ph.D. thesis, Cornell Univ., New York, U.S.A.
170. **Jewell, W. J., Carpener, H. R., and Dell'Oroto, S.,** Anaerobic fermentation of agricultural residue: potential for improvement and implementation, Final Report, HCP/T2981-07, NTIS, U.S.A., 1978.
171. **Atkinson, B., Daoud, I. B., and Williams, D. A.,** A theory for the biological film reactor, *Trans. Inst. Chem. Eng.,* 46, T245, 1968.
172. **Atkinson, B. and Davies, I. J.,** The overall rate of substrate uptake (reaction) by microbial films. I. A biological rate equation, *Trans. Inst. Chem. Eng.,* 52, 248, 1974.
173. **Atkinson, B., Swilley, E. L., Busch, A. W., and Williams, D. A.,** Kinetics, mass transfer and organism growth in a biological film reactor, *Trans. Inst. Chem. Eng.,* 45, T257, 1967.
174. **Atkinson, B. and Daoud, I. S.,** Diffusion effects within microbial films, *Trans. Inst. Chem. Eng.,* 48, T245, 1970.
175. **Atkinson, B. and How, S. Y.,** The overall rate of substrate uptake (reaction) by microbial films. Part II — Effect of concentration and thickness with mixed microbial films, *Trans. Inst. Chem. Eng.,* 52, 260, 1974.
176. **Monod, J.,** La technique du culture continue; theorie et application, *Ann. Inst. Pasteur,* 79, 390, 1950.
177. **Atkinson, B., Black, G. M., Lewis, P. J. S., and Pinches, A.,** Biological particles of given size, shape, and density for use in biological reactors, *Biotechnol. Bioeng.,* 21, 193, 1979.
178. **McCabe, W. L. and Smith, J. C.,** *Unit Operations of Chemical Engineering, 3rd ed.,* McGraw-Hill, New York, 1976, chaps. 7 and 9.
179. **Blanchard, J. P. and Gill, T. A.,** Suspended-particle, fixed biomass anaerobic digesters for livestock wastes, *Trans. ASAE,* 27(2), 535, 1984.
180. **Andrews, J. F., Cole, R. D., and Pearson, E. A.,** Kinetics and characteristics of multi-stage methane fermentations, S.E.R.L. Report 64-11, Univ. of California, Berkely, 1964.
181. **Feilden, N. E. H.,** The theory and practice of anaerobic digestion reactor design, *Process Biochem.,* 18(5), 34, 1983.
182. **Gates, W. E. and Marlar, J. T.,** Graphical analysis of batch culture data using the Monod expression. *J. Water Pollut. Control Fed.,* 40(11), R469, 1968.

183. **Pearson, E. A.,** Kinetics of biological treatment, in *Advances in Water Quality Improvement,* Gloyna, E. F. and Eckenfelder, W. W., Jr., Eds., Univ. of Texas Press, Austin, 1968, 381.
184. **Monod, J.,** The growth of bacterial culture, *Ann. Rev. Microbiol.,* 3, 371, 1949.
185. **O'Rourke, J. T.,** Kinetics of anaerobic treatment at reduced temperature, Ph. D. thesis, Stanford University, Stanford, Calif.
186. **Gates, W. E., Smith, J. H., Shun-Dar Lin, and Ris, C. H.,** An anaerobic-aerobic process for industrial wastewater treatment. I. The aerobic process, *J. Water Pollut. Control Fed.,* 39(12), 1951, 1967.
187. **Kornegay, B. H. and Andrews, J. F.,** Kinetics of fixed film biological reactors, in Proc. 22nd Ind. Waste Conf., Purdue Univ., Lafayette, Ind., 1967.
188. **Wehner, J. F. and Wilhern, R. H.,** Boundary conditions of flow reactors, *Chem. Eng. Sci.,* 6, 89, 1958.
189. **Haldane, J. B. S.,** *Enzymes,* Longmans, London, 1930, 84.
190. **Graef, S. P. and Andrews, J. F.,** Stability and control of anaerobic digestion, *J. Water Pollut. Control Fed.,* 46(4), 666, 1974.
191. **Andrews, J. F. and Graef, S. P.,** Dynamic modeling and simulation of the anaerobic digestion process, in *Anaerobic Biological Treatment Processes: Advances in Chemistry Series 105,* American Chemical Society, Washington, D.C., 1971, 126.
192. **Graef, S. P. and Andrews, J. F.,** Process stability and control strategies for the anaerobic digester, presented at the 45th Annual Conference, Water Pollution Control Federation, Atlanta, Georgia, October, 1972.
193. **te Boekhorst, R. H., Ogilvie, J. R., and Pos, J.,** A overview of current simulation models for an anaerobic digester, in *Livestock Waste: A Renewable Resource: Proceedings 4th International Symposium on Livestock Wastes,* ASAE, St. Joseph, Michigan, 1980, 105.
194. **Hill, D. T. and Barth, C. L.,** A dynamic mdoel for simulation of animal waste digestion, *J. Water Pollut. Control Fed.,* 49(10), 2129, 1977.
195. **Blanchard, J. P. and Gill, T. A.,** Design considerations for support particle suspension in a retained biomass reactor, *Trans. ASAE,* 28(1), 209, 1985.
196. **Perry, J. K., Ed.,** *Chemical Engineers Handbook,* 5th ed., McGraw-Hill, New York, 1973, chap. 19.
197. **Uhl, V. W. and Gray, J. B., Eds.,** *Mixing Theory and Practice,* Vol. 1, Academic Press, New York, 1966.
198. **Uhl, V. W. and Gray, J. B., Eds.,** *Mixing Theory and Practice,* Vol. 2, Academic Press, New York, 1966.
199. **Rushton, J. H., Costich, E. W., and Everett, H. J.,** Power characteristics of mixing impellers. I and II, *Chem. Eng. Prog.,* 46(8), 395 and 46(9), 467, 1950.

Chapter 3

FIXED FILM ANAEROBIC DIGESTION

Manuel J. T. Carrondo and M. Ascensão M. Reis

TABLE OF CONTENTS

I.   Introduction 102
    A.   Anaerobic vs. Aerobic Biotechnology 102
    B.   Fixed Film vs. Suspended Growth Reactors 103

II.   Microbial and Biochemical Fundamentals 103
    A.   Microbiology 104
    B.   Kinetics 104
    C.   Stoicheometry 105
    D.   Nutrient Requirements 105
    E.   Temperature 106
    F.   Inhibition and Toxicity 106

III.   Modeling the Kinetics of Anaerobic Biofilm 107
    A.   Hydrolysis of Solid Substrates 107
    B.   Soluble Substrate Removal in Biofilm 108
    C.   Out-diffusion of Products 110

IV.   Reactors and Processes 111
    A.   Processes 111
        1.   The Phased Process 112
    B.   Reactor Types 113
        1.   Fixed-Bed Reactor 113
        2.   Fluidized Bed Reactor 115
        3.   Expanded Bed Reactor 115
        4.   Rotating Bed Reactor 116
        5.   Recycled Bed Reactor 116
        6.   Recycled Flocs (Anaerobic Contact Reactor) 116
        7.   Upflow Anaerobic Sludge Blanket Reactor 116
    C.   Comparison of Reactor Types 117
        1.   Substrates 117
        2.   Toxic Substances 117
        3.   Foaming 119
        4.   Phasing 119
        5.   Recycle 119
        6.   Mixing 119
        7.   Overloading 119
        8.   Start-up 119
        9.   Gas Bubbles 119

V.   Design and Utilization 119
    A.   Traditional and Conceptual Design 119
    B.   Common Design Parameters and Relationships 120
    C.   Start-Up Recomendations 122
    D.   Wastes Amenable to Treatment by Fixed Film Anaerobic
        Reactors 123

E.   Full Scale Plants 124

VI.   Conclusions and Forecasts 125

Acknowledgments 126

References.. 126

I. INTRODUCTION

A. Anaerobic vs. Aerobic Biotechnology

Anaerobic digestion is becoming an important and attractive process for waste treatment as a result of a large research effort devoted to the development of the understanding of its fundamentals in the last few years, supported by significative advances in the process engineering aspects of anaerobic reactors.

The field of potential applications of anaerobic biotechnology in waste treatment is spreading fast; in many cases it has become an economic alternative to conventional aerobic treatment processes. The fact that almost any organic substance that can be aerobically degraded can also be so anaerobically,[1,2] has been important in providing such interest in anaerobic processes.

Some of the most salient features of anaerobic processes when compared to anaerobic process may be listed as follows.

Advantages of Anaerobic Treatment[3-5]

1.   Low production of stabilized sludge (0.04 to 0.14 kg sludge/kg COD removed, as compared to 0.3 to 0.5 kg sludge/kg COD removed in aerobic process).
2.   Possibility of recovering 80 to 90% of the energy present in the metabolized waste as methane gas (which may be used as a fuel).
3.   Low consumption of nutrients.
4.   Lower energy requirements than aerobic treatment (no need for aeration and smaller amounts of sludge for dewatering and disposal).
5.   High degree of waste stabiliztion (about 90%) compatible with very high organic loadings.
6.   Well acclimated anaerobic reactors can be preserved unfed for periods up to 1 year without any appreciable deterioration in sludge properties.
7.   Anaerobiosis sharply enhances public health quality of the wastes by inactivating pathogenic bacteria, as well as some viruses and parasites; when properly conducted no nuisant odor is produced.

Limitations of Anaerobic Treatment[3,4]

1.   Great sensibility to some specific compounds such as $CHCl_3$, CCl_4 and CN^- which have been found to be toxic.
2.   Due to the slow growth rate of the anaerobic bacteria the first start-up may take several months, although a period of 8 to 12 weeks is possible when innoculation with digested sewage is carried out.

3. A rather high temperature (30 to 40°C) is desirable to reach high conversion rates.
4. Anaerobic digestion is essentially a pretreatment method.

It has been estimated that anaerobic digestion offers a potential net energy saving of about 2.3×10^4 kJ/kg COD removed and that its operating costs are about \$90/ton* COD removed smaller than for aerobic treatment.[2]

In addition, studies conducted with an expanded bed reactor showed that the use of this anaerobic treatment may reduce capital costs from 40 to 80%, with installed energy requirements of only 10 to 20% of an equivalent aerobic system.[6]

B. Fixed-Film vs. Suspended Growth Reactors

The first reactors used in anaerobic treatment of domestic sewage sludge were of the completely mixed or the CSTR type. For these reactors, the hydraulic retention time is similar to the solids retention time (or sludge age).

As the anaerobic bacteria growth rates are slow, these reactors required high hydraulic retention times and consequently high reactor volumes for anaerobic digestion of low strength wastes, which resulted in an economically unfeasible reactor for large quantities of wastes.

One way to overcome this problem is to increase biomass retention time inside the reactor, either by creating inside the reactor conditions promoting natural biomass tendency to originate ''granulating'' flocs with enough size to be separed by settling and recycled to the reactor, or by introducing in the reactor a solid support where on the bacteria could attach and create a film that could be kept inside the reactor.[7]

In either case, hydraulic retention time can now be much smaller then the solids retentions time in these fixed film reactors. Using fixed-film reactor waste may be effectively treated in hours, as compared with several days for suspended growth.

The increased biomass retention time not only reduces digester volume and corresponding capital costs, but also allows methane production rates per unit of reactor volume 2 to 10 times greater than those reached with suspended growth reactors,[8] thus increasing their utilization for energy production.

Other advantages of fixed film compared to suspended growth reactors include less susceptibility to washout, higher tollerance to shock loadings, capacity to sustain active microbial culture even after periods of starvation and facility to withstand considerably higher transient and chronic concentrations of toxics.[9]

As in the past, suspended growth reactors are mainly used for concentrated wastes with an appreciable suspended solids content, while fixed-film reactors are more appropriate for the digestion of soluble low and high strength wastes;[10] the introduction of leachate types of digesters for solid wastes and of plug flow patterns and other types for wastes of high solid contents complicates the distinction made above but increases the potential use of anaerobic digestion.

II. MICROBIAL AND BIOCHEMICAL FUNDAMENTALS

Recent interest in anaerobic fermentation results from three main needs.

1. To utilize renewable resources for chemicals and fuels
2. To improve energy conservation in waste processes
3. To balance economic and political dependence on petroleum, which should be conserved for synthesis of higher molecular weight products that can only be produced from renewable resources at too high cost

* Dollars computed on U.S. 1986 standard.

Given the diversity of anaerobes their microbiology and biochemistry are only slowly being unravelled. Good recent reviews on the theme being available,[11-15] only a very short description of the main characteristics of importance to biochemical engineering of digesters will be considered here.

A. Microbiology

Generally anaerobic microorganisms can be classified as acidogenic, solventogenic (ethanol, butanol, or acetone producing), sulfidogenic and methanogenic. Methane being the major final reduced fermentation product, all the other species are potencially involved, as the same polymer can be degraded by acidogenic or solventogenic species and one- and two-carbon compounds can be fermented by methanogenic, acidogenic, or sulfidogenic species.[11,12]

Presently, four different trophic groups are known to exist in anaerobic digesters.[11-13,16]

1. Hydrolitic bacteria, catalyzing high molecular weight compounds (saccharides, proteins, and lipids)
2. Hydrogen producing acetogenic bacteria, fermenting fatty acids, and neutral end products
3. Homoacetogenic bacteria that can either catabolise unicarbon compounds (H_2/CO_2, HCOOH) or hydrolize multicarbon compounds to acetic acid
4. Methanogenic bacteria producing methane from acetate and one-carbon compounds

The sulfate reducers are known to interrelate with the methane producers, but the complex nature of this synthrophy is not yet fully understood, although since the discovery by Bryant,[17] of the syntrophic association of ''S'' organism and ''MOH'' organism in M. Omelianski, it has been known that pure culture of ''S'' organism are difficult to grow due to the accumulation of hydrogen; these sulfate reducing bacteria can grow in the absence of sulfur by utilizing the methanogenes as exogenous electron acceptors.[18]

It has now been established that propionate and longer fatty acids are catabolized by similar syntrophic associations.[19] The identifications of the rate contolling step depends mainly on the nature of substrate and to a lesser extent of the process configuration, temperature and loading rate.[1]

The rate limiting step will depend on the particular waste to be treated; generally for simple organic substrates rapidly fermented to volatile acids, conversion to methane will be the rate controlling step. When dealing with biopolymers like celluloses, the hydrolitic step prior to acid fermentation will be rate limiting, this situation will worsen in the presence of a recalcitrant compound such as lignin.[13]

B. Kinetics

Even if in fixed film reactors microbial kinetics are rarely limiting, a knowledge of the order of magnitude of the intrinsic kinetic paramenters (assuming Monod's approach) is in order. Since metabolic activities of pure cultures very often do not correlate well with those ocurring in mixed populations, much of the microbiology information on pure culture and pure substrates should be used with caution. On the other hand, when utilizing mixed populations and real wastes, the range for the kinetic parameters is expectedly very large as can be seen in Henze and Harramoes review.[20]

Thus, although these should be used for preliminary estimates only, average values can be indicated for the maximum specific growth rate (θ max, day^{-1}) of 1.5, 0.3 and 0.3 for the acid producing, methane producing bacteria and combined cultures, respectively and, similarly, for the half velocity constant (ks, kg COD m^{-3}) at 0.2 and 0.05 for acid and methane producers.[20] Although thermodynamically fundamental papers[14,15] indicate a larger diversity in interpreting energy yields based on ATP production, the traditional value of 10 g cells formed per mole ATP generated[21,22] used to calculate the maximum growth yields is reasonable and thus "averages" of 0.15, 0.03, and 0.18 for Y_{max} (kg VSS/kg COD) for acid producing, methane producing, and combined culture can be indicated; the observed yield coefficients decrease as the organic load decreases and thus, for the combined culture may be as low as 0.04 kg VSS/kg COD,[20,23] further decreasing the "sludge produced" to be disposed of in a digester and increasing methane yield (its maximum being 0.37 m³-STP/KgCOD degraded at a bacteria yield coefficient of zero).

For the same temperature as previously considered, 35°C, the maximum substrate removal rates r_s (μ_{max}/Y_{max}, kg COD/kg VSS/day) can now be estimated, depending on the percentage of the active biomass; since this is normally lower than 60% VSS, r_s of 1 kg COD/kg VSS/ day for combined cultures are practical maximums consistent with the literature,[20] although the "theoretical maximum" at 100% active VSS is 1.7 kg COD/kg VSS/day. Caution must be exerted in using these removal rates when suspended solids are already present in the influent — these must then be accounted for in an overall mass balance correction.

C. Stoicheometry

The knowledge of the yield constants above, coupled with traditional chemical information allows the stoicheometry of the anaerobic process to be established for methane production, alkalinity changes, nutrient requirements, and even toxicants like sulfide.[24,25]

Less elaborate predictions of biomass composition and quality can be obtained using the Buswell-Mueller formula, applicable to the main classes of organic materials based on the knowledge of feed chemical composition as follows.

$$C_nH_aO_b + \left(n - \frac{1}{4}a - \frac{1}{2}b\right) H_2O \rightarrow \left(\frac{1}{2}n - \frac{1}{8}a + \frac{1}{4}b\right) CO_2 + \left(\frac{1}{2}n + \frac{1}{8}a - \frac{1}{4}b\right) CH_4$$

Pratically, maximal yields are 443 ℓ CH₄/kg cellulose 370 ℓ/kg sugar, 620 ℓ/kg lipids, and 493 ℓ/kg proteins. Finally, biogas composition can also be asserted from the composition of organic material as COD and TOC as follows.[26]

$$COD/TOC = V(CH_4)/(V(CH_4) + V(CO_2)) \times 5.33$$

The theoretical compositions thus, predicted are modified by alcalinity, dilution rate, and temperature. Whereas methane is almost insoluble and unreactive in aqueous phase, part of the carbon dioxide produced is dissolved in the liquid phase and, mainly with the ammonia provided by protein degradation, enhances the biocarbonate buffering effect; this leads to an enrichment of methane in the gas phase further enhanced by higher dillution rates removing more carbon dioxide in the liquid phase. Apart from other microbiological effects, an increase in temperature acts in the opposite way by reducing carbon dioxide solubility.

D. Nutrient Requirements

The bacteria responsible for waste conversion and stabilization in anaerobic processes require optimal nutrient supply in order to achieve optimum growth.

Although some wastes, namely municipal wastewater sludges, usually contain all the adequate nutrients, in other cases the addition of nutrients may be necessary.

For the two larger nutrients next to carbon substrate, Speece and McCarty,[27] established the phosphorus requirement to be about 15% of the nitrogen requirement, the amount of each being determined by cell synthesis; the COD/N ratio decreases with an increase in organic load, with a theoretical minimum of approximately 50:1. Other micronutrients, such as iron, cobalt, nickel, and sulfide, have been shown to be obligatory for methanogens to convert acetate to methane.[27] The trace amounts required are dependent on the particular type of waste to be digested.

Given the higher solids retention times and higher loading rates of fixed film systems, these can normally be operated with smaller additions of nutrients than equivalent suspended growth reactors.[1]

E. Temperature

Anaerobic digestion can take place within each of the temperature ranges at which microorganisms can operate: thermophilic (50 to 70°C); mesophilic (30 to 45°C) and psychrophilic (10 to 25°C).[28] The choice of optimum depends on economic and operational factors and the most common anaerobic processes are operated at mesophilic temperature.

Advantages and disadvantages of working at the thermophilic range seem to be slightly controversial,[13,20,26] but general features can be summarized as follows.

1. Higher metabolic rates than with mesophilic temperatures thus permitting shorter retention times — this is even true for hydrolytic bacteria
2. Above 60°C, acetoclastic methanogens seem to be inhibited,[26] though other microbiologists quote an increase in methanogenesis up to 65°C[13]
3. There is a more limited species diversity at thermophilic temperature and this is probably responsible for less stability in operation[20] this again being somewhat controversial[26]
4. Lower growth yields resulting in slower start-up and smaller capability to accommodate shock loads or toxics.

The psychrophilic range is quite stable[29,30] but as rates are low it is economically less interesting unless very unsophisticated systems are desirable.

F. Inhibition and Toxicity

In digesters, methanogenic bacteria are the most sensitive to toxic substances. The presence of toxicants may cause inhibition of methanogens with concomitant decrease on methane production and eventually process failure. Often this decrease in methane production is accompanied by an increase in organic acids.[31] Traditionally it has been considered that concentration of volatile acids should not exceed 500 mg ℓ^{-1},[32] the undissociated acids acting as the inhibitor and propionic acid being more toxic than acetic. Nevertheless, concentrations up to 5 g/ℓ have recently been suggested as possible if buffering capacity exists to keep the pH above 7.[26] Some toxicants, namely heavy metals like nickel and copper are needed for metabolism at low concentrations and a similar situation takes place with sulfide; the simultaneous presence of heavy metals and sulfide tend to cancel each other out as the insoluble sulfides precipitate. Sulfide toxicity is developing as an operational problem as more and more anaerobic digesters became operational. The graph presented by Speece can be very useful.[1]

On the other hand, small concentrations of sulfide stimulate methanogenesis as it acts as a sulfur source for biosynthesis and as source of reducing power to keep the redox potential (Eh) below -300 mv, value at which methanogenesis has been known to proceed effectively.[33]

Owen et al.[34] presented a relatively simple and inexpensive technique for measuring the toxicity of constituents in feed sources to anaerobic treatment. This anaerobic toxicity assay (ATA) measures the adverse effect of a compound on the rate of the total gas production from an easily utilized methanogenic substrate;[34] this biossay can also be used to determine the concentration at which particular compounds exhibit toxicity and the period required for a microbial population to acclimate to them.[35]

Many other toxicants are reported in the literature, namely, furfural, formaldehyde, chloroform, cyanide, phenol, etc. For some of them, long term microbial acclimation is known to lead to degradation in anaerobic processes.[36]

Finally it should be mentioned that long solids retention time (SRT), as provided by the use of attached growth reactors, reduce toxicity problems.[9]

III. MODELING THE KINETICS OF ANAEROBIC BIOFILM

Kinetical modeling in biotechnology as elsewhere in chemical engineering, serves the dual purpose of allowing deeper, conceptual verification of the theories behind the model and in its mathematical form, permits better design, control, and operation of the reactors.

Usual biochemical engineering approaches generally consider both biochemical reaction and mass transfer mechanisms, either in lumped (i.e., series) or distributed (i.e., parallel — both reaction and transfer are occurring everywhere in the pellet or floc film) models.[37] The first are usually simpler to manipulate mathematically although often the second approach is required for realistic description. In this case, effectiveness factors, Thiele modulus, and saturation parameters similar to those utilized in heterogeneous catalysis, are used.[38] Generally, pure culture and simple substrates are considered and a limiting substrate can be identified.

In wastewater treatment a few additional problems arise as

1.   The complex substrates are normally described by "global" parameters like chemical or biochemical oxygen demand (COD;BOD) that do not distinguish between undegraded products and metabolites.
2.   The mixed cultures utilized have different kinetical parameters and interact in multiple ways.
3.   On top of the metabolic degradation pollution removal by biosorption is also possible.[39]

With these extra difficulties, biofilm kinetical modeling has also been developed for wastewater treatment, mainly for aerobic[40] or anoxic (nitrification-denitrification) conditions[41,42] with models for specifically anaerobic reactors being much more scarce.[43] Two main approaches have been used, similar to those applied in biochemical engineering: the lumped model usually considers a global reaction rate, similar to chemical kinetics, be it zero, first, or half order, depending on biofilm thickness; the distributed models will depend on more kinetical and diffusional parameters as Monod's kinetics are used in a more complex manner thus being further away from practical application.

A brief discription follows on the kinetics for the hydrolysis of solid substrates, soluble substrate removal in the biofilm and the out diffusion of products from the biofilm. It should be realized that not only is this description very introductory but also that, namely for anaerobic processes, much of what has been suggested/proposed has not been thoroughly tested.

A. Hydrolysis of Solid Substrates

Eastman and Ferguson,[45] proposed a kinetic model applicable to the hydrolysis of particulate substrates, assuming that hydrolysis rate (R_h) at constant temperature and pH is directly proportional to the concentration of degradable particulate COD (F) with a first order hydrolysis rate constant (K_h):

$$R_h = K_h \cdot F$$

From suspended culture experiments values of the first order hydrolysis rate constant of $K_h = 0.125$ hr^{-1} at 35°C and pH $= 5.5$ were obtained.[45] It is also known that lowering the pH increases hydrolysis rate,[47] but no kinetical description is known to the authors. It is also not known whether in a biofilm reactor similar laws or constants to those suggested above apply; as exocellular enzymes are responsible for much of the hydrolysis process, an influence is expected from the use of different innocula.

Once solubilized, the substrate can diffuse into the biofilm where it is degraded to acetic acid or methane at the acid phase or methane phase respectively.

B. Soluble Substrate Removal in Biofilm

Substrate removal from an aqueous phase by a biofilm requires:[41]

1. Diffusion of the substrate into the biofilm
2. Metabolism by bacteria
3. Out diffusion of the metabolic products through the biofilm and into the aqueous phase

The molecular diffusion process follows Fick's first law. The rate of reaction in the biofilm can be considered as o'-order 1'-order, or Monod's kinetics.

De Walle and Chian,[48] indicated that the substrate removal rate is affected by substrate concentration, specific surface area, flow rate, and temperature. These authors experimentally achieved removal rates of 7 kg COD/(m^3 day) and K_s values of 1.5 g/ℓ.

Harramoës[44] presented a simple model applicable to anaerobic processes, capable of predicting the fraction of methanogenic and acidogenic species in available biofilm space, with or without diffusional limitation. The methanogenic fraction can be written as follows:

$$\alpha_m = \frac{Y_m\, r_{A,m}}{Y_m r_{A,m} + Y_a r_{A,a}}$$

where: $\alpha m =$ fraction of methanogenic bacteria in the biofilm; $Y =$ yield constant; $r_A =$ removal rate per unit surface; m $=$ methane phase; and a $=$ acid phase.

For a specific o'-order model this fraction depends on the following dimentionless quantities:

$$\frac{Y_m\, \mu_m}{Y_a\, \mu_a} \quad \text{and} \quad \frac{Ds\, Ss}{Do\, So}$$

where $\mu =$ specific growth rate constant; $S =$ substrate concentration in the bulk liquid; $D =$ diffusion coefficient in the biofilm; o $=$ stand for the substrate for acid phase; and s $=$ stand for the substrate for methane phase.

The same authors determined the biofilm thickness, beyond which the diffusional resistance becomes important for anaerobioses for a bulk concentration of 0.4 kg COD/m^3, concluding that diffusional resistance has no practical significance for biofilm thickness lesser than 1 mm.[20]

An example of a distributed model was presented by Rittmann and McCarty,[49] derived from the biofilm model developed by Williamson and McCarty,[41] and providing approximate analytical solutions.

Based on the assumption of the validity of Monod's kinetics, this is a variable order model, the global order of reaction varying regularly as a function of influent organic loading; it includes liquid layer mass transport, biofilm molecular diffusion, Monod's kinetics, biofilm growth and decay, and assumes that required nutrients exist in excess but for one which is the rate limiting substrate.

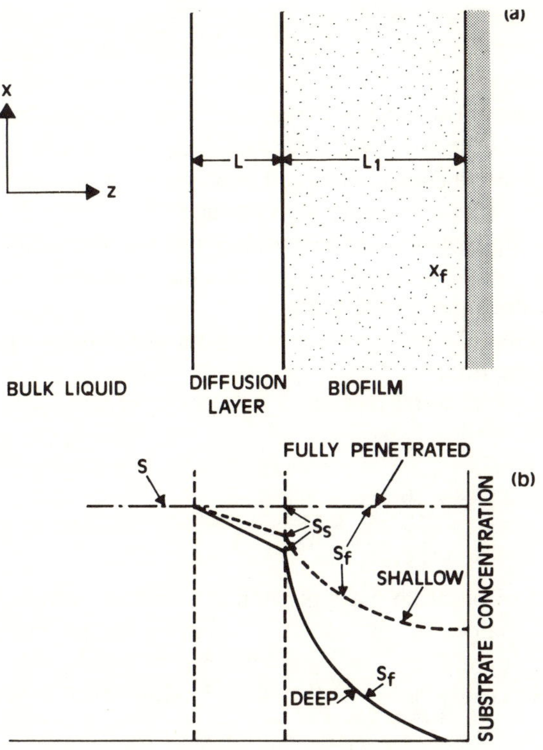

FIGURE 1.   Conceptual basis for biofilm model: (a) physical concepts; (b) substrate concentration profiles.[43]

The fundamentals of the variable order model are presented in Figure 1. The idealized biofilm has an uniform thickness and an uniform cell density of X_f. The substrate is transported unidimensionally to the biofilm surface and through the biofilm.

Similarly to Harremoës,[44] three characteristic substrate concentration profiles are considered:

1. A deep biofilm in which S_f (the substrate concentration in the biofilm) decreases assymptotically to zero.
2. The fully penetrated biofilm where S_s (the substrate concentration at the liquid biofilm interface) is constant throughout the full thickness.
3. The intermediate situation, where $O < S_f < S_s$ called shallow biofilm.[43]

The combination of Fick's law of molecular diffusion and Monod's kinetics leads to the usual nonlinear differential equation allowing for the determination of the steady state substrate concentration within a differential section of biofilm:

$$D_f \frac{d^2 \, S_f}{dz^2} = \frac{K \cdot X_f \, S_f}{K_s + S_f}$$

where: D_f = molecular diffusivity in the biofilm; K = maximum specific rate of substrate utilization; and Ks = half velocity constant.

To predict transport resistance from the bulk liquid to the biofilm surface across the diffusion layer, Fick's first law can be applied to yield:

$$\frac{J}{A} = \frac{D(S - S_s)}{L} = K_m(S - S_s) \quad \text{with} \quad K_m = \frac{D}{L}$$

where: A = surface area normal to the z direction; J = substrate flux into the biofilm; D = molecular diffusivity of the substrate in the liquid; and K_m = mass transport coefficient.

The resolution of these equations with eight independent variables (S, L, D, Df, K_s, k, X_f, and L_f) is carried out in the dimensionless domain.[49]

The same authors simplified the model by coupling substrate utilization to the biofilm growth for a steadystate biofilm defined as one that has neither net growth nor decay over time (i.e., the biofilm thickness is constant).[43]

This model assumes that the total amount of biofilm mass is just equal to that which can be supported by substrate flux.

The net growth rate of bacterial mass in a differential section of biofilm is represented as:

$$\frac{\partial \, dz}{\partial \, t} = Y \, \frac{kS_f}{K_s + S_f} \, dz - bdz$$

where: Y = true yield of bacterial mass per unit of substrate utilized; b = specific decay or maintenance respiration coefficient; and dz = thickness of the biofilm section.

The steady state biofilm thickness can be determined by equating growth due to substrate flux (JY) with the maintenance decay of the entire biofilm (bX_fL_f) thus:

$$L_f = JY/b \, X_f$$

This equation requires that the steady-state biofilm be dynamic, i.e., that there is a balance between the growing and the decaying sections of the biofilm such that its thickness remains constant; as no further losses (due to predation, sloughing, or exchange with the bulk liquid) are considered, the thickness thus calculated may be an over estimate.

Based on kinetic and energetic constraints this model predicts that a steady-state bulk concentration, Smin, exists below which no significant biofilm activity should occur. Equations are available to calculate the steady state substrate flux and biofilm thickness for S > Smin.[43]

This model was applied to different reactor configurations by Rittmann,[50] to determine the expected performance on steady-state conditions. As expected, this shows that simple loading factors and kinetic relationships are insufficient to describe biofilm processes and also explains why fluidized-beds perform better than completely mixed and fixed-bed reactor.

Various limitations rectrict the applicability of this model namely the accurance of situations like nonsteady-state conditions, dual substrate limitations, wall effects, bed expansion, media clogging, and shear losses, which may account for substantial deviations of the observed results from those predicted by the model.[50]

In fluidized-bed anaerobic reactors using molasses as substrate, biofilm thickness is always under 100 μm and the reactions are zero order.[51] Given the low cellular yields typical of anaerobiosis, this might also be the case for other reactor types, thus substantially simplifying the kinetical description of the systems — such an hypothesis is currently under investigation in our group for expanded bed, upflow sludge blanket and anaerobic filter, for which the hypothesis is rather less probable.

C. Out-Diffusion of Products

The out-diffusion of the products formed inside biofilms may have an important role in the entire process of substrate removal by biofilms, mainly where the diffusional resistance is important to the biofilm process.

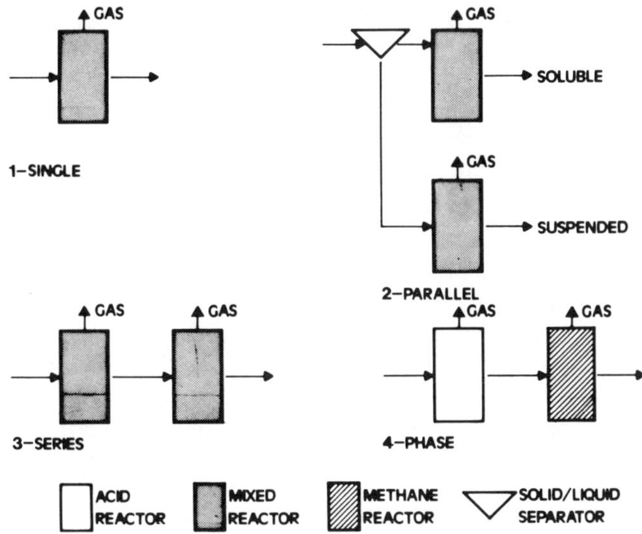

FIGURE 2.   Anaerobic processes.

Nevertheless the majority of the models so far proposed to simulate the performance of biofilm reactors do not explicitly take into account the role played by the diffusional resistance in the out-diffusion of the products.

Some experimental observations suggest that diffusional resistance of products formed may be the explanation of certain phenomena occuring in biofilms that are otherwise difficult to explain, and that in certain circumstances it may even result in an alteration of the biofilm kinetics. This might be the case with the increase of the pH in the biofilm with the increase in the organic carbon species or with the methane bubble formation in the biolfim.[20]

IV. REACTORS AND PROCESSES

There is a wide variety of systems and reactor types presently used for anaerobic treatment; however, a closer look reveals that some of these are in fact different designations for fundamentally the same reactor or process.

In this section the main features of basic types of processes and reactors will be presented, with a view to enable the reader to chose the best system for a given wastewater treatment problem. A small summary of fixed-film full-scale plants known by the authors ends the section.

A. Processes

Anaerobic digestion is carried out inside one reactor or combination of reactors, so that there are four major different processual schemes (Figure 2).

Single anaerobic process — This is the conventional process, consisting of only one reactor where both acid and methane phases occur. The combination of more than one of these reactors, connected in series or parallel, originates the staged or the parallel process respectively.

Parallel Anaerobic Process — In this processing scheme, the wastewater to be treated is first separated into a soluble phase and a suspended phase in a solid liquid separator (generally a sedimentation tank), which are then separately treated in two or more independent reactors, comprising both the acid and the methane phase.

Staged Anaerobic Process — This scheme consists of two or more reactors connected in series, each of them performing both the acid and methane phase of anaerobic digestion.

Phased Anaerobic Process — In this phase process hydrolysis and acidogenesis is dominant in the first stage "acidogenic" digester, aceticlastic methanogenesis being dominant in the second "Methanogenic" digester; the production of methane from carbon dioxide through the hydrogenotrophic methanogens takes place in both reactors, as it is faster than the aceticlastic methanogens.

In deciding which process layout to use for a given wastewater, it is useful to consider the concept of solubility index introduced by Henze and Harremoes.[20] Wastewater with high solubility index (0.8 to 1.0) mainly consisting of soluble matter, or low solubility index (less than 0.2) mainly consisting of suspended solids, maybe treated either in a single reactor or in a staged or phased process in a single process line.[20] For medium solubility index between 0.2 to 0.8 the parallel process may be advantageous as it separates the soluble from the suspended phase which can then be treated in the most appropriate type of reactor.

By combining these four basic processes it is possible to obtain several different layouts suitable for treating wastewaters of different solubility indexes and different compositions.[20] The two phase anaerobic process has potential advantages over the conventional process and so it will be covered in more detail in the following section.

1. The Phased Process

As stated above, the first phase reactor will mainly produce acids, after the occurrence of hydrolysis of polymers whereas the second phased reactor will mainly produce methane. The acetogenic conversion of the higher carbon acids, being also a slow reaction may take place mainly in the second reactor if highly biodegradable substrates, rapidly fermented to higher acids and hydrogen by fermentative bacteria, constitute the majority of the feed; acetogenic conversion will take place mainly in the first reactor if it is charged with particulate substrates, especially if high sludge and hydraulic retention times are utilized for more efficient hydrolysis.

Generally the "two bacterial groups" involved in these steps differ significantly with respect to cell physiology, nutrient requirements, growth rate, and optimum environmental conditions.[52,53] For instance the optimum pH value for the first group of bacteria is about 5 to 6, while for the methanogenic bacteria is 7 to 8.[20,46,54] This creates a difficult control problem for single reactor with respect to pH.

Furthermore, from the nutritional point of view, methanogens will operate from metabolites of the acidogenic organisms and their growth rate (mainly for aceticlastic) are much smaller than for acidogenes (the acetogens being more of an exception as refered above), so different washout rates will exist, thus allowing more economical dimensioning of the reactors in the two phase process.[46,52,53]

For these reasons Ghosh et al. proposed to run each phase in separated reactors in what has been termed the phased process.[47] In the presence of more solid substrates, needing to undergo hydrolysis, the lower pH possible in the first reactor helps, as the rate of hydrolysis increases as the pH value decreases.[46,47] The main advantages of phase separation are summarized as follows.

1. It allows an independent monitoring and control of some operational parameters such as pH, temperature, oxidation-reduction potential (ORP), biomass recycle, hydraulic, and solids retention time, etc., for each phase.[47,53,55,56]
2. Substantial reduction in total reactor volume, (and resulting lower capital costs) can be obtained, as the reactor sizes for each phase are based on the different growth kinetics peculiar to each group of bacteria.[47]
3. Increased biomass activity is possible. A biomass activity of 1.5 kg COD/(kg VSS day) was reported from methane forming bacteria in a two phase process, compared with a more usual value of 0.5 kg COD/(kg VSS day) for single process.[58]

4. Improved capacity for treating toxic substances. Sulfate or other toxic substances present in wastes such as those from the pulp and paper industry may be confined to the first reactor, decreasing the inhibition effect on the more sensible methanogenic organisms in the methane reactor.[20,30,57]
5. Better overall performance is shown by the two phase process than the single process even if longer retention times are used for this one.[47,53]

The two phase separation has been claimed to reduce digestion instability caused by the accumulation of hydrogen, propionic, and butyric acid;[58] it might also allow the removal from the acid reactor of chemicals produced.[54,59,60]

The main drawback of phased operation is the need for skilled operation and for additional instrumentation for monitoring and control,[47] although operational control becomes easier as stated above. Phase separation is usually operated by kinetic control, (short residence time), chemical control (low pH value or addition of methanogenic inhibitors), or a combination or both.[46,47,53-55,61]

Process optimization for phased processes depends on the type of substrates used. For less soluble substrates the first phase should be operated in such a way to maximize the hydrolysis rate (limiting step for these type of substrates). For soluble substrates hydrolysis is not critical and attention should be directed towards the quality of the intermediate species produced in the first phase as substrates for methanogenesis.[53]

A process similar to phase is that used by Rijkens,[62] and Institute of Gas Technology,[63] shortly designated by leaching bed leachate filter. This process also consists of two reactors, the first producing the solubilization of the solid feedstock, while in the second fermentation to methane occurs.

B. Reactor Types

This section will focus on fixed-film reactors. The general characteristics of the reactors were refered in Section I,C.

Basically seven different fixed film reactor types may be considered, as shown in Figure 3. It should be pointed out that two of the reactor types indicated in Figure 3, the upflow sludge blanket (UASB), and the recycled flocs reactor, are usually operated as high solids retention time suspended growth reactors becoming fixed-film reactors only if floc granulation occurs during operation.[20,29]

1. Fixed-Bed Reactor

A typical fixed-bed is a column reactor filled up to 75 to 100% with an inert material such as gravel rocks, coke, or some manufactured plastic or ceramic media. The microorganisms become attached to the media surface or are flocculated in the interstitial spaces, bacteria washout being thus diminished. The amount of retained biomass depends on many factors including the surface/volume ratio and the attached growth is limited by the suport area,[64] there is no need for mixing nor in most cases, for effluent recirculation and thus this reactor presents a large plug flow pattern, the wastewater passing through the bed either up- or downflow; recirculation might be needed if some control of biofilm thickness, toxicity or pH is envisaged.

Young and McCarty were the pioneers in the developement of fixed-bed anaerobic reactors in the late 1950s.[29] Since that time increasing experience has been gained with this type of reactor as laboratory, pilot, and full-scale plants have shown that the anaerobic filter is suitable for treating various types of wastes with high efficiencies. The main features of the anaerobic filter are as follows.

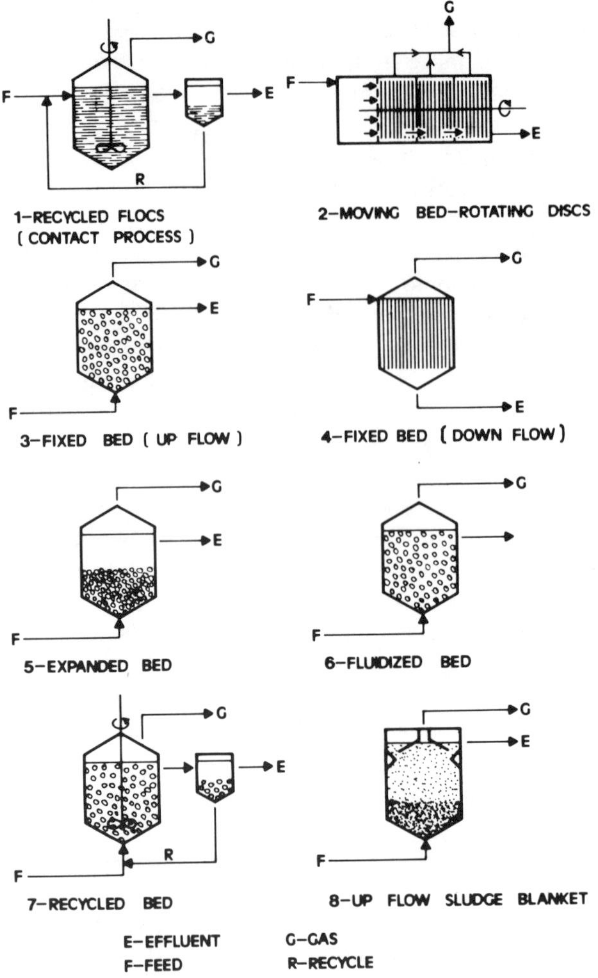

FIGURE 3. Anaerobic reactor types.

1. It is capable of handling high organic loading,[65] being particularly suited for treatment at psychrophilic temperatures (10 to 25°C).[29]
2. It is suitable for operation at variable loading rates and under-shock loads;[65] loads above 90 kg COD/m³ day for over a 24 hr period without the performance being affected has been quoted.[66]
3. It can start up very quickly after a period of starvation (1 or 2 days to reach the maximum capacity after 3 weeks of starvation).[64,65]
4. It is capable of operation at short or very short hydraulic residence times (lows of about 10 min were reported).[7]

The principal disadvantage of the anaerobic filter is the high cost of filter media, (which increases the capital costs of reactor) and the need for periodic backwashing. Backwashing is necessary mainly when treating high strength wastes or wastes containing influent suspended solids that are not biodegradable. Due to the accumulation of large quantities of biological solids clogging might occur consequently decreasing the reactor efficiency.[10] Excess solids removal can also be achieved by partially draining the suspended solids traped in-between the inert media.[7]

To minimize clogging problems and to deal with wastes containing appreciable quantities of suspended solids, Van den Berg and Lentz developed a downflow stationary fixed-film reactor in which the filling material consists of oriented straight vertical channels.[67] Similar downflow stationary bed reactors have been operated at high organic loads of suspended solids and short retention time,[64] or at ambient temperatures.[68,69]

2. Fluidized Bed Reactor

In this reactor type the microorganisms are attached to an inert media, e.g., sand, activated carbon, or gravel.[20] The support media is fluidized by a high upwards flow of the wastewater obtained by recycling a great portion of the effluent, and volume expansions of about 30 to 100% are usually achieved. Due to the high vertical velocity, in the range 2 to 20 m hr^{-1},[70] each particle is moved around the bed generally located within a small bed volume,[20,71,72] this plus the bed regeneration and expansion strategy and size and density of the inert in relation to the upflow rate control the biofilm thickness.[20,71] The fluidization of the small media results in extremely large specific surface area (2500 m^2/m^3 has been reported) and concomitantly high biomass concentration (10 to 20 kg VSS/m^3).[20,71-73]

The size of support media influences the recycle ratio required for a given expansion of the bed (ratios of 5 to 500 were reported);[20,74] when sand particle size of 0.7 mm is used, twice the recycle rate is required than for a particle size of 0.5 mm, in order to achieve a 30 to 100% bed expansion.[20,75] Thus, the use of small and lighter particles will reduce the recycle ratio and consequently will reduce the pumping costs which represent the major disadvantage of this process; the use of particles with a diameter below 0.1 mm has been suggested.[74] Other features of the fluidized-bed reactor include[73,75]

1. Maximum contact between liquid and media.
2. Minimal liquid film diffusional resistances due to the biofilm thickness control.
3. Problems of channeling, plugging, gas holdup, and high pressure drop are avoided.
4. High recycle rates help diluting high strength wastes.
5. Suitable for large scale operation.
6. It has an uniform solids distribution.

COD removal efficiencies in excess of 80% are possible operating a fluidized-bed reactor for treatment of wastes with high carbohydrate content at loading rates greater than 16 kg COD/m^3 day;[75] at the other extreme this process is said to be also suitable for treatment of very dilute wastes (100 to 1000 g COD/m^3 were reported).[76]

3. Expanded Bed Reactor

As with the fluidized-bed, similar inert materials are also used as support, the expanded bed reactor differing from the fluidized-bed process in the degree of bed expansion used, ranging from 10 to 20%.[77] Again, the bed expansion is controlled by the recycle ratio, which is maintained low (20 to 60),[78] to reduce upflow velocities in the reactor in the range of 2 to 10 m/hr,[20] just sufficient to prevent clogging of bed. Due to the low upflow velocity, the particles keep their place within the bed, the biofilm thickness being controlled by physical contact between them.[20]

The main features of this reactor type are similar to those indicated for the fluidized-bed; the process efficiency being also partly due to the large surface area-to-volume ratio (1000 to 3000 m^2/m^3),[20] enables high biomass concentration (up to 30 kg/m^3) within the reactor.[79]

Much of the original work on the expanded bed reactor was carried out at Cornell University, by Jewell and Switzenbaum.[79] The expanded bed process is suitable for medium strength (1.5 to 3 g COD/ℓ) and high strength (5 to 16 g COD/ℓ) soluble wastes, achieving a total COD removal efficiency of 70% at a volumetric organic loading rate as high as 30 kg COD/(m^3 day) at thermophilic temperature.[80] More recently, this process was applied to the digestion of pure cellulose in laboratory scale, a cellulose conversion of about 90% being obtained at 30°C and at loading rates up to 8 kg COD/(m^3 day).[78] It has also been used to treat dilute industrial wastewaters (<1 kg COD/m^3) and domestic wastewater,[77] with good removal rates, although the economics of its use for this purpose are not as well established.

4. Rotating Bed Reactor

This reactor type, also named anaerobic rotating biological contactor (A$_n$ RBC), consists of a series of discs (generally packed in three or four "stages"), mounted on a horizontal shaft within a cylindrical tank; these discs are immersed in the wastewater continuously rotating and providing mixing action. Each stage is separated by baffles to prevent short-circuiting, biomass being attached to the surfaces of the discs.[81] Excess sludge leaves the reactor together with the treated wastewater. Scarce information exists, but it has been applied to high strength wastewaters (8.5 kg COD/m^3).[81,82]

5. Recycled Bed Reactor

This reactor type consists of a mixed tank containing approximately 3% (v/v) of a support material (sand, anthracite, etc.). A substantial part of the biomass is found in suspended flocs which are kept in suspension by mechanical or gas mixing, solids leaving the tank from an external settler.[83] Excess sludge in the reactors can be drained from the bed recycling pipe. This reactor type has been named by Mattersson and Frostell,[83] as Carrier Assisted Sludge Bed Reactor (CASBER).

The major problem associated with this reactor type is the separation of solids from the wastewater in the external settler. This drawback may be overcome by gas stripping or cooling of the influent to the separator,[20] neither solution being very practical or by enclosing the settler on top of the bed.

The CASBER has been used for treating beet sugar factory waste of 4 to 7 kg COD/m^3 and synthetic molasses wastewater of 9.3 kg COD/m^3, organic loading of 25 kg COD/(m^3 day) and 5 to 6 kg COD/(m^3 day), respectively, being tolerated.[83]

6. Recycled Flocs (Anaerobic Contact Reactor)

The recycled flocs (anaerobic contact reactor) consists of a mixed tank without purposeful addition of support material and as such is usually a high solids retention time suspended growth reactor. Nevertheless the function of support material may be provided by inert particles existing in wastewater to be treated, or by granulation of the flocs and thus it will become similar to the recycled bed just described.

As with the recycle bed reactor, the main problem with this type of reactor is the separation and concentration of biomass flocs in the settler and similar solutions to the recycled bed have been suggested: gas stripping or cooling of the influent to the separator,[20] or enclosing the settler on top of the bed.

The use of this reactor type for wastes with relatively low inert concentration (either organic or inorganic suspended solids) has been described.[84]

7. Upflow Anaerobic Sludge Blanket Reactor

The upflow anaerobic sludge blanket (UASB) reactor, was developed by Lettinga and co-workers,[3] and is conceptually similar to the recycled bed. This reactor type is based on upflow movement of wastewater through a dense blanket of solids consisting of microorganisms and unreacted solids,[56] and expanded by the liquid upflow and gas bubbling.

The reactor is equipped in the upper part with a system for gas-liquid-solid separation, providing an adequate sludge retention time.[85]

Within the UASB a highly settleable biomass develops as flocs or as granules (1 to 5 mm). Usually a dense bed of granules develops at the bottom, while a flocculant biomass develops near the gas liquid surface, thus increasing a solids profile along the reactor.[3] The granular sludge has superior settleability characteristics over flocculated sludges, especially if the later is rich in filamentous bacteria.[3]

Organic loading rates of 15 to 40 kg COD/(m³ day) and up to 16 kg COD/(m³ day) were handled when the operation of the UASB in pilot and full-scale plant, respectively, for operation in the mesophilic range.[3] Several full-scales plants are already in operation with sizes ranging from 100 m³ up to 5000 m³, treating a wide range of wastes such as sugar beet waste, potato processing waste, potato starch waste, maize starch waste, shell fish wastes, and brewery waste.[85]

This process is feasible for acidogenic step in anaerobic treatment being then capable of handling organic loading rates of 60 to 80 kg COD/(m³ day).[3] For high solids content wastes (up to 8%) Ghosh[63] utilizes a similar type of reactor for an acidogenic and methanogenic phase.

Similar reactor to the UASB was called tower digester by Callender et al. at Sidney University,[7] the principal difference to the UASB being that chemical focculants were added to enhance the formation of biomass flocs with high velocities.

C. Comparison of Reactor Types

As indicated at the beginning of this section, a difference was established between fixed-film and suspended growth or anaerobic flocs based on attachment. Their performance will also differ, partly because in fixed-film anaerobic reactors the retained biomass depends mainly in surface to volume ratio being thus limited by material support area, whereas in anaerobic floc reactors it is limited by the settling rate of the particles and the fluid properties of concentrated biomass suspension.[64]

Since conceptually some of the fixed-film reactors are similar to anaerobic floc types and these, if granulation (mineralization and densification of the inner part of the flocs) occurs, became fundamentally fixed-film reactors as the ''granules'' then constitute an inert attachment, differentiating between them is not always easy and is not of major importance. A brief review of some of the differences in behavior between the various types follows as well as a tabular comparison (Table 1) and a pictorial description (Figure 3).

1. Substrates

For different wastes to be anaerobically treated different reactor choices are available to be used;[83] for high solids content wastes the recycled bed and the recycled flocs,[63] or its variations and the upflow sludge blanket[87] are the most appropriate, although as pointed above, an expanded bed reactor has been used to degrade cellulolytic wastes;[78] the fixed-bed (downflow)[67] and rotating bed reactors,[81] can be used with medium to low suspended solids concentration, whereas generally the upflow fixed-bed,[29,30] and the expanded[80] and fluidized-bed reactors,[88] are suitable for soluble wastes.

2. Toxic Substances

It has been indicated that the anaerobic filter, anaerobic fluidized/expanded bed, anaerobic upflow sludge blanket and rotating bed reactors are the most effective types for treating industrial wastewaters containing toxicants.[2] In drastic toxic loads, Norrman reports that during exposure to high concentration of toxic substances the expanded bed reactor was upset the biofilm washed away, while in the anaerobic filter only the gas production was stopped; the filter recovered while the expanded bed had to be reinoculated.[89]

Table 1
COMPARISON BETWEEN ANAEROBIC REACTOR TYPES

Characteristics (units)	Fixed bed	Rotating bed	Expanded bed	Fluidized bed	Recycled bed	Recycled flocs	Sludge blanket
Diameter of support material (mm)	20—50	1000—3000	0.3—3	0.2—1	0.01—0.1	—	—
Specific surface area (m^2/m^3)	60—200	100—200	1000—3000	1000—2500	2000—5000	—	—
Vertical empty bed velocity (m/hr) (recycle)	0.01—0.10	—	2—10	6—20	—	—	0.05—0.30
Biomass concentration (kg SS/m^3)	5—15	5—15	10—20	10—30	5—15	—	5—15
Attached biomass (% of total)	20—80[a] 50—90[b]	50—80	90—100	95—100	60—80	—	60—80
Suspended biomass (% of total)	20—80[a] 10—50[b]	20—50	0—10	0—5	20—40	—	20—40
Energy consumption (Wh/m^3)	20—40	25—100	30—1000	80—3000	15—50	—	20—60
COD loading rate (kg/m^3 day)	1—10[a] 5—15[b]	—	1—20	1—20	—	1—6	5—30
COD removal (%)	80—95[a] 75—88[b]	—	80—87	80—87	—	80—95	85—95
Susceptibility to toxic substances	No	No	Yes	Yes	Yes	Yes	Yes
Problems with foaming	No	No	Yes	Yes	Partially	Partially	Yes
Problems with gas bubbles	Partially	No	Partially	Partially	No	No	Partially
Recycle necessary	No	No	Yes	Yes	Yes	Yes	No
Mixing necessary	No	No	No	No	Yes	Yes	Partially
Tolerates organic overloading	Yes	Yes	Partially	Partially	Yes	Yes	Partially
Start-up problems	Partially	Partially	Yes	Yes	Partially	Partially	Yes

Note: Modified from Henze and Harremoes[20] and including data from References 3, 29, 64, 71, 79, 83 to 86, and 90.

[a] Upflow.
[b] Downflow.

3. Foaming

Although all reactor types have foaming problems it seems that fixed and moving bed reactor are less susceptible than other reactors.[20] Foaming in the reactor may be prevented by addition of an antifoaming agent to the feed solution, by stirring the top with scum breakers or by spraying the top intermittently with the liquid.

4. Phasing

In general all reactor types can be operated within a phasing process. However when fixed and moving bed type are used as methanogenic reactors, recycling might be needed to control pH problems.[20]

5. Recycle

The expanded and fluidized-bed reactors require recycle to maintain the bed in suspension increasing substantially their operation costs. The recycled bed and recycled flocs reactor have lower degree of recirculation than the formers. For the other reactor types recycle might only be necessary to control toxicity and pH problems.[9]

6. Mixing

Where needed, mixing can be accomplished by mechanical stirring, gas mixing or by recycling of wastewater. The recycled bed and recycled flocs reactor need some sort of mechanical stirring in order to keep the bed and flocs suspended.[83]

Lettinga et al.[3] indicated that in UASB, the mechanical mixing will only be necessary if low organic loads (2 to 4 kg $COD/(m^3$ day)) and high hydraulic loads (3 to 4 $m^3/(m^3$ day)) are applied and in treating wastes containing significant amounts of undissolved solids.

7. Overloading

The recycled bed, recycled flocs, and sludge blanket reactors are more susceptible to washout when subjected to an hydraulic overload than the other reactor types,[20,64] although operational difficulties might lead to bed losses in the fluidized-bed. Reactor types where performance depends on biofilm/flocs characteristics such as fluidized-bed, expanded bed and sludge blanket, are more sensitive to overloadings, which tend to alter the biofilm or flocs characteristics.[20]

8. Start-Up

As previously indicated, the main problem of the anaerobic fixed film reactors relates to start-up difficulties. Generally, expanded and fluidized-bed, plus UASB are believed to give more problems than the other reactor types,[20,31] fixed bed being the easiest to start-up.[64,85] Some general recomendations for start-up are presented in Section V,C.

9. Gas Bubbles

Gas holdup in the form of gas bubbles sometimes cause difficulties with fixed-film anaerobic digesters, namely expanded and fluidized-bed, and sludge blanket reactors. Adherence of biomass particles or flocs to the bubbles may result in biomass washout; furthermore rising bubbles may cause channeling and the clogging of the bed due to entrapping, causing a reduction in the effective reactor volume.[20,90]

V. DESIGN AND UTILIZATION

A. Traditional and Conceptual Design

The design of fixed-film anaerobic reactors has predominantly been based on empiric criteria and/or previous experimental work. Most designs use the results from pilot plants and thus the variation in the parameters chosen for design as well as their relationships is not surprising.

On the other hand, with the increasing popularity and widespread utilization of biofilm reactors, the need for unified and theoretically developed methods is an important task. These models should be firmly supported by a good understanding of the fundamentals of anaerobic fixed-film processes and capable of adequately predicting the performance of the several types of reactors and of being useful for their design.

A brief overview of the basic kinetical models has been presented in Section III. Here a short description will be made for the two basic approaches used in the design of anaerobic fixed-film processes, the traditional-empirical design and a more conceptual design. Both approaches converge in attempting to achieve a reactor design meeting the required specification for any given wastewater treatment problem, performing efficiently and economically its function. However, due to the complexity of the anaerobic fixed-film process no clearly established methods have been found for this purpose.

The traditional approach bases the design of waste treatment reactors on some practical parameters and its empirical relationships with the purification desired.[20] Usually, data from pilot plant or laboratory tests are translated into practical design figures, typically loading, and removal parameters, such as these referred to in Table 2.

The use of these loading and removal rates generally gives acceptable results if applied within the range of conditions of conventional experience, problems arise when trying to extrapolate outside this range.[20] Therefore, the generalization of these parameters as design criteria is limited, as important features of the process are ignored.

Another problem with these empiric design criteria for biofilm reactors is the inconsistency among them. For instance the reactor volume considered for volumetric loading and detention time is sometimes taken as empty-bed volume and sometimes as void volume for the bed. The same happens with surface area and hydraulic loads that have been based either on the biofilm area and/or the crosssectional area normal to flow direction.[50] Admittedly, some standardization would remove this problem.

Finally, it should be noticed that the merit of this approach relies precisely on its simplicity, avoiding the need for highly sophisticated models with complicated mathematical solutions.

The conceptual approach attempts a more complete understanding of the processes involved in order to produce models capable of adequately simulating them and predicting the purification results for a given waste.[20] Therefore these models try to solve a conflict: to be as general as possible and stated in such a way that their applications in design should be fairly easy. As an example of the efforts in developing models capable of general application to the design of different types of biofilm reactors is the utilization of an unified model for the design of five basic types of biofilm reactors, demonstrating that simple criteria such as loading rates cannot predict the performance of biofilm processes.[50]

On the other hand this and most other similar models involve sophisticated nonlinear mathematics or the use of difficult to measure parameters resulting in limited practical application.[91] Thus, a unified conceptual design for anaerobic fixed-film reactors is needed. Meanwhile, the traditional-empirical approach will be the workhorse of the designer.

B. Common Design Parameters and Relationships

In the design of anaerobic digestion reactors using either or both of the design approaches above mentioned a number of different parameters need to be evaluated for good reactor performance.

As indicated by Feilden,[92] two objectives in anaerobic digester design are trying to get a fast growth rate of the bacteria and avoiding biomass washout with the liquid stream. Additionally, other objectives may include optimization of methane production, minimization of energy requirements and obtention of good removal rates.

Table 2
COMMON DESIGN PARAMETERS AND RELATIONSHIPS

Parameter	Relationships	Dimensions
Mass flow rate to the reactor	$F = Q.C_1$	$M_i T^{-1}$
Total sludge mass in reactor	$V.X$	M_x
Total media surface in reactor	$V.a$	L^2
Mean hydraulic residence time	$t = V/Q$	T
Mean solids retention time	$\theta_x = V.X/Q(X_1 - X_2)$	T
Superficial velocity	$V = Q/A$	$L\,T^{-1}$
Hydraulic loading per unit surface media	$Q/V.a$	$L\,T^{-1}$
Volumetric loading rate	$B_v = F/V$	$M_i\,L^{-3}T^{-1}$
Mass loading rate per unit sludge mass	$B_x = F/V.X$	$M_i\,M_x^{-1}T^{-1}$
Mass loading rate per unit cross-sectional area	$N_c = F/A$	$M_i\,L^{-2}T^{-1}$
Mass loading rate per unit surface media	$B_A = F/V.a$	$M_i\,L^{-2}T^{-1}$
Substrate removal per unit volume (empty bed)	$r_v = Q(C_1 - C_2)/V$	$M_i\,L^{-3}T^{-1}$
Substrate removal per unit surface media	$r_A = Q(C_1 - C_2)V.a$	$M_i\,L^{-2}T^{-1}$
Substrate removal per unit sludge mass	$r_x = Q(C_1 - C_2)/V.X$	$M_i\,M_x^{-1}T^{-1}$

Note: Q = volumetric flow rate ($L^3\,T^{-1}$); C_1 = total substrate concentration in the influent ($Mi\,L^{-3}$); C_2 = total substrate concentration in the effluent ($Mi\,L^{-3}$); X = Suspended solids concentration in the reactor ($M_x\,L^{-3}$); X_2 = suspended solids concentration in the effluent ($M_x\,L^{-3}$); V = volume of reactor (empty bed) (L^3); A = cross sectional area (empty bed) (L^2); and a = specific interfacial area (L^{-1} (or L^2L^{-3})).

Some of the design parameters and their definitions are indicated in Table 2; wherever possible the notation proposed by the IAWPRC/IUPAC working group has been used;[93] Table 1 appearing in Section IV,C, lists ranges for some of these parameters and different reactor types.

Among the parameters referred above, volumetric loading rate defined as the substrate loading per unit of volume per day, is probably the most used in design. It can be related to the suspended solids concentration in the reactor and the mean solids retention time by the following equation[20]

$$\mathrm{Bv,\ COD} = \frac{(X/\theta_x - B_v,\ \mathrm{inert})}{Y}$$

Figure 4 shows the dependence of the volumetric load on the yield coefficient, for different biomass concentrations for acid, and methane reactor. From this figure and the above equation several conclusions can be drawn.

1.   For the same yield coefficient the maximum volumetric loading rate is obtained for the larger biomass concentration.
2.   For the normal ranges usually encountered inerts concentration has a small influence in the maximum volumetric loading.
3.   Wastes with low yield coefficient allow much higher volumetric loadings than those with higher yield coefficients.
4.   For the same volumetric loading, the higher yield coefficient is obtained for higher biomass concentration.
5.   When operating phase separated processes, larger loading rates can be applied to the acid reactor than to the methane reactor.

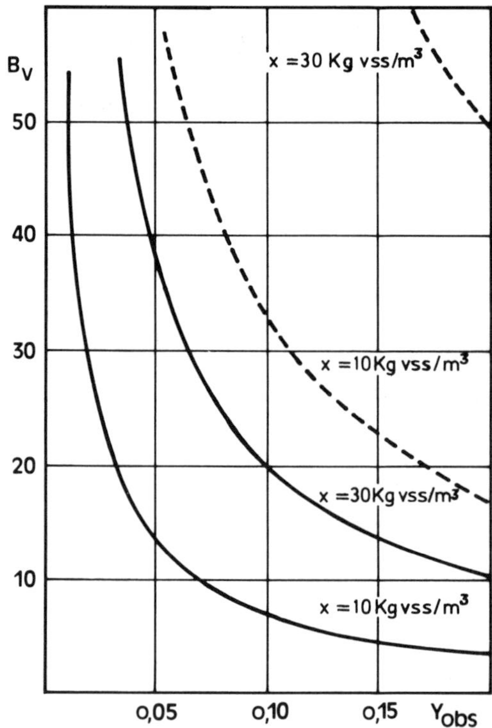

FIGURE 4. Volumetric load as a function of the observed yield coefficient at 35°C and $B_{v\ inert} = 0$ for acid production ($- -$, $\theta_x = 3$ days) and methane production (——, $\theta_x = 15$ days).[20]

C. Start-Up Recommendations

As previously mentioned the start-up of an anaerobic process is often the most difficult stage of the entire operation procedure as it will also influence later process stability. Not much written information is available as the more basic knowledge is lacking and some of what is available is conflicting. Thus, the next few paragraphs summarize the most important points according to the literature and the authors experience, and should be used as a guideline for further improvement. As in the other anaerobic processes, start-up of fixed-film process should take into consideration factors such as the type of waste, the type of innoculum to be used, the reactor design, and the inert support material used. Factors such as the pH, alkalinity, biomass washout, organic loadings, and volatile fatty acids should be closely controlled in order to reach as soon as possible steady-state operation at the desired conditions.

It is well known that the generation of the convenient type of microorganisms to the waste to be treated can be achieved by an innoculum addition to the reactor (seeding), resulting in an acceleration of start-up period.[3,29,69,80,94] For this purpose the innoculum can be sludge of a municipal digester, manure, soil, etc.,[20] but the start-up period is considerably faster (as much as four times) when sludge from an anaerobic process treating identical substrates is used,[29,69] sludge with good settling properties enhancing the start-up of UASB reactor.[3] The recommended quantity of innoculum varies according to different researchers. Young and McCarty found that the most effective way of starting the anaerobic filter was to add heavy seed only to the bottom third of the reactor,[29] whereas Hobson et al.[94] state that the greater volume of an active innoculum used the quicker and more surely digestion will start. As a rule it seems that the innoculum quantity should be more than 10% of the reactor volume, with the optimum value varying between 30 and 50%.[95]

For recycled flocs reactors operating at high organic loadings Van den Berg used a continuous innoculation procedure and this could also be advantageous during start-up periods for other type of reactors.[96] In order to provide for a rapid adaptation of the innoculum to the waste to be treated, the additions of adequate nutrients to the feed, particularly if there is a low content of any of these in the waste, might be performed.[13]

Also to be considered is the type of support media used as the degree of attachment of the bacteria biofilm to the support is also dependent on it. To be a good support for biofilms, a material must have a porous structure. In this sense the granular activated carbon or porous polymers yield better result than sand.[95]

A comparison between materials with different porosities has been carried out showing that the use of less porous materials such as glass or pvc results in longer start-up periods than with fired clay support;[69] for plastics there might also be some problems with potentially toxic plasticizers.[97] Good attachment seems to be facilitated if the microorganisms exhibit a sticky surface or if it produces glue-like substances such as extracellular polymers.[95] Since this is not the case with the majority of anaerobic microorganisms, the provision of nutrients for their growth has been suggested as a help to obtain a good attachment.[95] Dextran, molasses, sugar as well as polyacrilamides operating as flocculants themselves, have been quoted for this purpose.[95] Alternatively the generation of a slime matrix around the support media can be obtained by a previous aerobic operation before anaerobic operation starts.[95]

Temperature choice is important during the start-up period and it is recomended to start-up at 35°C, even when the process will later operate at different temperature.[95] pH and alkalinity must also be kept at the optimum range for the type of microorganism to be used.

Biomass washout must be reduced to the minimum during start-up; this can be accomplished either by a proper control of operational parameters (e.g., HRT, organic loading) or by using any device capable of solid-liquid separation and recycling biomass. H.R.T. must be kept at a high level during the initial period of start-up and organic loading should not exceed initial values of 0.1 kg COD/(kg VSS/day).[3] As gas production increases the organic loading can gradually be increased. Martenson and Frostell,[83] operating a recycled bed reactor, report that only after 1 month was it possible to increase the organic loading. In fixed-bed reactors, increases in organic loading between 25 and 100%/week were preferred.[98]

Volatile fatty acids must also be monitored during start-up and if increased levels are observed (> 0.5 kg HAC/m³) the organic loading must be reduced until the concentration of volatile fatty acids decreases.[3,20]

Lettinga et al.[3] presented some specific recomendations concerning the start-up of an UASB.

1. The amount of seed sludge must be 10 to 20 kg VSS/m³.
2. The initial mass loading rate per unit sludge mass must be 0.05 to 0.1 kg COD/(kg VSS day).
3. Volumetric loading rate should not be increased until all fatty acids are well degraded (80% for medium strength wastes).
4. The environmental conditions for the growth of the anaerobic bacteria should be favorable.

Even under these specification the start-up period of an UASB will take 6 to 12 weeks.[3]

D. Wastes Amenable to Treatment by Fixed Film Anaerobic Reactors

From the early days, when anaerobic treatment was generally considered for municipal wastewater sludges only, a large change has occurred, mainly over the last 10 to 15 years. Today, almost any residue that is amenable to aerobic treatment can also be treated anaerobically; many wastes that are hard to treat aerobically have been shown to be degradable anaerobically.[36]

Table 3
**SOME WASTES AMENABLE TO ANAEROBIC
BIOTECHNOLOGY**

Amenable Wastes

Animal wastes	Municipal wastes and refuse
Beef cattle waste	Solid refuse
Dairy cow manure	Municipal landfill effluent
Cattle manure	Primary sludge
Swine manure	Activated sludge
Piggery wastes	Municipal refuse
Beef cattle manure	Other wastes
Livestock manure	Water hyacinth
Broiler-chicken litter	Straw
Milk waste	Bermuda grass
Sloughterhouse	Kelp
Calf manure	Wine distillery wastes
Fruit and vegetable wastes	Rum stillage
Spinach waste	Distillery
Asparagus peels	Cider
French bean waste	Brewery
Apple pulp	Soft drink bottling
Apple slurry	Chemical
Carrot waste	Petrochemical
Green pea slurry	Pharmaceutical
Been blanching wastes	Leachates from landfills
Potato peeling waste	Sulfite evaporator condensate
Potato blanching	Kraft mill
Molasses	Black liquor condensate
Vegetable tanning	Sulfite liquor
Maize starch	Sulfite pulping
Wheat starch	Paper pulping
Palm oil	Paper board
Olive oil	Pulp
Sauerkraut	Wood fiber
Whey permeate	Corn stover
Whey	Pectin wastes
Sugar beet plant	Peat
Soy processing	

Nevertheless and although municipal wastewaters can be treated by fixed-film reactors,[77] it is generally recognized that from an economic viewpoint the anaerobic process only became attractive at waste concentrations above 1 to 2 kg COD/m^3.[99]

Table 3 is just a short list of wastes for which experimental data exists in using fixed film anaerobic digestion with potential economic benefits given the fact that environmental laws have to be obeyed.

E. Full-Scale Plants

For the reasons already mentioned, cheaper wastewater treatment with the advantage of energy production in the form of biogas and the possibility of utilizing the sludge as fertilizer, anaerobic digestion is now spreading fast.

Nyns et al.[100] reported that in a period of 5 years, ending in 1983, at least 550 anaerobic digesters for methane production were installed in Europe and Switzerland alone. Only a small part of the plants that are presently built, are designed according to proprietary processes, most of them being designed and built without references to any specific patented process. A list of proprietary processes and installations known to the authors where those

Table 4
ANAEROBIC BIOTECHNOLOGY PROPRIETARY INSTALLATIONS

Company	Process name	No. of installations	Reactor type
AC-Biotechnics (ex-sorigona) (Sweden)	Anamet	U.S. (3) Canada (1) Europe 19	Anaerobic contact[a]
Dorr-Oliver (U.S.)	Anitron	U.S. (2)	Fixed-film fluidized-bed
Esmil-csm (Netherlands) and Joseph Oat (U.S.)	Biothane	U.S. (3) Europe (18)	Upflow sludge blanket
Bacardi (U.S.)	Bacardi	U.S.(1)	Downflow packed bed
Badger (Celanese) (U.S.)	Celrobic	U.S.(3)	Upflow packed bed
Biomechanics (U.K.)	Bioenergy	Europe(3)	Anaerobic contact[a]
Ecolotrol (U.S.)	Hy-fio	U.S.(1)	Fixed-film fluidized-bed
Iris (France)	—	Europe(7)	Sludge bed anaerobic contact[a]
Sulzer (Switzerland)	Anoder	—	Upflow sludge blanket[a]
Enso-gutzeit (Finland)	Enso-fenox	Europe(2)	Fixed-film fluidized-bed
Autotrol (U.S.)	—	—	Moving bed (rotating biological contactor)
Air liquide (France)	—	France(2)	Upflow fixed bed
SGN (France)	—	France(2)[b]	Downflow fixed bed
CAPPA (Gist Brocades) (Netherlands)	—	Europe(2)[b]	Fixed-film fluidized-bed

[a]   Some of these reactors may not be considered fixed-film unless floc granulation occurs (see introduction).
[b]   One under construction (end 1984).

have been used is shown in Table 4 totaling 70 plants. The biggest is the Bacardi fixed bed with a total of 12,800 m^3 and SGN in now completing the construction of a 5000 m^3 downflow fixed-film bed digester in the Cognac Region, in France.

VI. CONCLUSIONS AND FORECASTS

The relationships between this "environmental" utilization and broader biotechnological development are becoming clearer. For high volume low value added products, like ethanol, biotechnology used (and this is still the largest market utilization), suspended growth, generally batch-although continuous operation is rapidly coming on stream for large ethanol producing plants in, e.g., the U.S. and Brazil, reactors with concentrations of cells (dry matter) of the order of 20 g/ℓ,[101] similar to those used in CSTR classical digesters. To decrease reactor size increasing conversion rates and operational flexibility immobilized cell reactors have been studied, allowing for concentrations of the catalyst ranging up to 100 g dry cell/ℓ reactor,[102,103] two to three times larger than those obtained in fixed film digesters, admittedly the cheapest form of immobilization (see Table 1). To further increase the cell concentration, membrane separated reactors with cell recycle are now undergoing evaluation, cell concentrations of 150 g/ℓ having been used (as compared to a theoretical maximum of approximately 200 g/ℓ if one considers cells as being made up of approximately 80% water);[104] given the fact that methane is the lowest value added product one can think of in biotechnological terms, the complexity of substrates — wastewaters used and the large cost of membranes or products for immobilization — it is not very probable that they will be used in wastewater treatment by anaerobic digestion but the possibility cannot be entirely ruled out and might be worth investigating.

From a different angle, some of the knowledge gained in biotechnology in running solid substrate fermentation and coping with its mass transfer problems[105] may improve the potential for utilization of "dry anaerobic digestion" as carried out by Jewell et al. at Cornell[106] or even improve the operation of "engineered sanitary landfills" for biogas production.

The increased knowledge[117,118] in cellulose utilization pretreatment — enzymatic,[109] acid,[110] and alkaline[111] — might also be applied to more solid feeds. Here, once pretreatment costs have been taken into account, methane will not be the economic choice but possibly just the acid obtained in the acid phase digestion — be it as salts, esters, or ketones[60] — if cheap separation procedures (liquid-liquid extraction)[61,112,113] ion exchange, or membranes,[114] can be made to work. Knowledge of the optimization of acetobutyric fermentation will be needed as well as single or engineered co-culture techniques to increase yields; among others, our laboratory is looking at this possibility under a contract with the Program in Science and Technology Cooperation, Washington, U.S.A.

Other areas that will have to be improved include circumvention or removal of inhibitory toxicants, as in biotechnology, even though some aspects will be more "wastewater" specific like the sulfide problem (also under investigation at our laboratory) well known in molasses slops,[30] or from farm wastes, in more developed countries,[115] as well as work on other toxicants, like those arising from pulp and paper, as mentioned before (Section II,F).

As these reactors become larger and more complex and process failures need to be avoided, more rugged instrumentation for monitoring will be developed and coupled with the outcome of modeling efforts to permit effective process control — another major aim of biotechnology.

Thus, even though the field is already ripe enough for many commercial operations to exist there is still much work to be carried out at all levels of research, development, and engineering before it will be a mature field and, as pointed out here, much of this work will run parallel to what is now taking place in the broader and encompassing field of biotechnology.

ACKNOWLEDGMENTS

The authors acknowledge the financial support of the Program in Science and Technology Cooperation, U.S.A.I.D., Washington, D.C. under Grant 936-5542-G-SS-4003-00. That permitted to extend the insight into the field as well as the continuation of the work in this area in our laboratory.

REFERENCES

1. **Speece, R. E.,** Anaerobic biochemistry for industrial wastewater treatment, *Environ. Sci. Technol.,* 17(9), 416A, 1983.
2. **Parkin, G. F., Speece, R. E., Yang, C. H. J., and Kocher, W. M.,** Response of methane fermentation systems to industrial toxicants, *J. Water Pollut. Control Fed.,* 55(1), 44, 1983.

3. **Lettinga, G., van Velsen, A. F. M., Hobma, S. W., de Zeeuw, W., and Klapwijk, A.,** Use of the upflow sludge blanket (UASB) reactor concept for biological wastewater treatment especially for anaerobic treatment, *Biotechnol. Bioeng.,* 22, 699, 1980.

4. **Frostell, B.,** Biological treatment of paper industry wastewater combined with energy recovery. Paper presented at Sweden's International Pulp and Paper Week, Stöckholm, 1981.

5. **Forday, W. and Greenfield, P. F.,** Anaerobic digestion, *Effluent Water Treat. J.,* 10, 405, 1983.

6. **Barnes, D., Bliss, P. J., Grauer, R. B., Kuo, C. H., and Robins, K.,** Pretreatment of high strength wastewaters by an anaerobic fluidized bed process. I. Overall performance, *Environ. Technol. Lett.,* 5(5), 195, 1983.

7. **Callender, I. J. and Bardford, J. P.,** Recent advances in anaerobic digestion technology, *Process Biochem.,* 8, 24, 1983.

8. **Jewell, W. J. and Mckenzie, S. E.,** Microbial yield dependence on dissolved oxygen in suspended and attached systems, in *Proc. 6th Water Resources Symp.,* University of Texas, Austin, 1972.

9. **Parkin, G. F. and Speece, R. E.,** Attached vs. suspended growth anaerobic reactors: response to toxic substances. In anaerobic treatment of wastewater in fixed film reactors, *Water Sci. Technol.,* 15, 261, 1983.

10. **Colleran, E., Barry, M., Wilkie, A., and Newell, P. J.,** Anaerobic digestion of agricultural wastes using the upflow anaerobic filter design, *Process Biochem.,* 17(2), 12, 1982.

11. **Zeikus, J. G.,** Chemical and fuel production by anaerobic bacteria, *Ann. Rev. Microbiol.,* 34, 423, 1980.

12. **Zeikus, J. G.,** Metabolic Communication between biodegradative populations in nature, in *Microbes in Their Natural Environmental,* Symp. 34, Slater, J. M., Whittenbunny, R., and Wimpenny, W. T., Eds., Society for General Microbiology Ltd, University Press, Cambridge, 1983, 423.

13. **Zeikus, J. G.,** Microbial population in digesters, in *Proc. First Int. Symp. on Anaerobic Digestion,* Stafford, D., Ed., A. D. Scientific Press, Cardiff, U.K., 1980, 75.

14. **Thauer, R. K. and Morris, J. G.,** Metabolism of chemotrophic anaerobes: old views and new aspects in the microbe 1984. II. Prokaryotes and Eukaryotes, in *Proc. Society for Microbiology,* Symp., 36, Kelly, D. P. and Carr, N. G., University Press, Cambridge, 1984.

15. **Thauer, R. K., Jungermann, K., and Decker, K.,** Energy conservation in chemotrophic anaerobic bacteria, *Bacteriol. Rev.,* 41, 100, 1977.

16. **McInerney, M. Y. and Bryant, M. P.,** Syntrophic associations of H_2 utilizing methenogenic bacteria and H_2-producing alcohol and fatty acid degrading bacteria in anaerobic degradation of organic matter, in *Anaerobes and Anaerobic Injections,* Gottschalk, H., Ed., Gustav Fisher Verlag, Stuttgart, 1980, 117.

17. **Bryant, M. P., Wolin, E. A., Wolin, M. J., and Wolfe, R. S.,** *M. Omellanski* a simbiotic association of two species of bacteria, *Arch. Mikrobiol.,* 59, 20, 1967.

18. **Le Gall, J., Moura, J. G., Peck, H. D., Jr., and Xavier, A. V.,** Hydrogenase and other iron-sulphur proteins from sulfate-reducing and methane forming bacteria, in *Iron Sulfur Protein,* Spiro, T. G. Ed., John Wiley & Sons, New York, 1982, 177.

19. **Boone, D. R. and Bryant, M. P.,** Propionate-degrading bacterium *Syntrophobacter Wolinii* s.p. nov., gen. nov., from methanogenic ecosystems, *Appl. Environ. Microbiol.,* 40, 626, 1980.

20. **Henze, M. and Harramoes, P.,** Review paper in anaerobic treatment of wastewater in fixed film reactors, *Water Sci. Technol.,* 15, 1, 1983.

21. **Bauchop, T. and Elsden, S. R.,** The growth of microorganisms in relation to their energy supply, *J. Gen. Microbiol.,* 23, 457, 1960.

22. **McCarty, P. L.,** Energetics of organic matter degradation, in *Water Pollution Microbiology,* Mitchell, Ed., Wiley-Interscience, New York, 1972, 91.

23. **Frostell, B.,** Anaerobic treatment in a sludge bed system compared with a filter system, *J. Water Pollut. Control Fed.,* 53, 216, 1981.

24. **Graef, S. P. and Andrews, J. F.,** Mathematical modeling and control of anaerobic digestion, *CEP Symp.,* 70(136), 101, 1974.

25. **Ferguson, J. F., Eis, B. J., and Benjamin, M. M.,** Neutralization in anaerobic treatment of an acid waste, Proc. (IAWPR) Specialised Sem: Anaerobic Treatment of Wastewater in Fixed Film Reactors, Copenhagen, 1982.

26. **Dubourguier, H. C., Albagnac, G., and Verrier, D.,** Methane production processes by fermentation of biomass, *CEE ISPAA Courses,* Reidel Publ., Synfuels, 1985.

27. **Speece, R. E. and McCarty, P. L.,** Nutrient requirements and biological solids accumulation in anaerobic digestion, in *Advances in Water Pollution Research,* Vol. 2, Eckenfelder, W. W., Ed. Pergamon Press, Elmsford, N.Y., 1964, 305.

28. **Bailey, J. E. and Ollis, D. F.,** *Biochemical Engineering Fundamentals,* McGraw-Hill, New York, 1977, 344.

29. **Young, J. C. and McCarty, P. L.,** The anaerobic filter for waste treatment, *J. Water Pollut. Control Fed.,* 41, R160, 1969.

30. **Carrondo, M., Silva, J., Ganho, R., Oliveira, J., and Figueira, M. I.,** Anaerobic filter treatment of molasses fermentation wastewater, *Water Sci. Technol.,* 15, 117, 1982.

31. **Andrews, J. F.,** Control Strategies for anaerobic digestion processes. II., *Water Sewage Works.,* 122, 74, 1975.

32. **Anderson, G. K., Donnelly, T., and McKeown, K. J.,** Identification and Control of inhibition in the anaerobic treatment of industrial wastewater, *Process Biochem.,* 17, 28, 1982.

33. **MacGregor, A. N. and Keeney, D. R.,** Methane formation by lake sediments during in vitro incubation, *Water Resour. Bull.,* 9, 1153, 1973.

34. **Owen, W. F., Stukey, D. C., Healy, J. B., Jr., Young, L. J., and McCarty, P. L.,** Bioassay for monitoring biochemical methane potential and anaerobic toxicity, *Water Res.,* 13, 485, 1979.

35. **Benjamin, M. M., Woods, S. L., and Ferguson, J. F.,** Toxicity and biodegradability of pulp mill waste constituents to anaerobic bacteria, Proc. of IAWPR: Specialised Sem: Anaerobic Treatment of Wastewaters in Fixed Film Reactors, Copenhagen, 1982.

36. **Suidan, M. T., Cross, W. H., and Fong, M.,** Continuous bioregeneration of granular actived carbon during the anaerobic degradation of catechol, *Prog. Water Technol.,* 12, 203, 1984.

37. **Bailey, J. E. and Ollis, D. F.,** *Biochemical Engineering Fundamentals,* McGraw-Hill, New York, 1977, 389.

38. **Satterfield, C. N.,** *Mass Transfer in Heterogenous Catalysis,* MIT Press, Cambridge, 1970.

39. **Roques, H.,** Ou en est la modelisation des processus de deppollution biologique, *Environ. Technol. Lett.,* 1, 151, 1980.

40. **La Motta, E. J.,** Internal diffusion and reaction in biological films, *Environ. Sci. Technol.,* 10, 1651, 1976.

41. **Williamson, K. and McCarty, P. L.,** Model of substrate utilization by bacterial films, *J. Water Pollut. Control Fed.,* 48, 9, 1976.

42. **Harramöes, P.,** The significance of pore diffusion to filter denitrification, *J. Water Pollut. Control Fed.,* 48, 337, 1976.

43. **Rittmann, B. E. and McCarty, P. L.,** Model of steady state biofilm kinetics, *Biotechnol. Bioeng.,* 22, 2343, 1980.

44. **Harramöes, P.,** Biofilm Kinetics, in *Water Pollution Microbiology,* Vol. 2, Mitchell, R., Ed., John Wiley & Sons, New York, 1978, 71.

45. **Eastman, J. A. and Ferguson, J. F.,** Solubilization of particulate organic carbon during the acid phase of anaerobic digestion, *J. Water Pollut. Control Fed.,* 53, 353, 1981.

46. **Cohen, A., Zoetmeyer, R. J., Andel, J. G., and Van Deursen, A.,** Anaerobic digestion of glucose with separated acid production and methane formation, *Water Res.,* 13, 571, 1979.

47. **Ghosh, S., Conrad, J. R., and Klass, K. L.,** Anaerobic acidogenesis of wastewater sludge, *J. Water Pollut. Control Fed.,* 47, 30, 1975.

48. **De Walle, F. B. and Chian, E. S. K.,** Kinetics of substrate removal in a completely mixed anaerobic filter, *Biotechnol. Bioeng.,* 18, 1275, 1976.

49. **Rittmann, B. E. and McCarty, P. L.,** Variable order model of bacterial film kinetics, *J. Environ. Eng. Div.,* 104(EE5), 889, 1978.

50. **Rittmann, B. E.,** Comparative performance of biofilm reactor types, *Biotechnol. Bioeng.,* 124, 1341, 1982.

51. **Elmaleh, S.,** Personal communication, 1984.

52. **Pohland, F. E. and Ghosh, S.,** Anaerobic stabilization of organic wastes: two phase concept, *Environ. Lett.,* 1, 255, 1971.

53. **Verstraete, W. and de Braene, L.,** Phase separation in anaerobic digestion: motives and methods, *Trib. Cebedeau,* 453(34), 367, 1981.

54. **Zoetemeyer, R. J., Van den Heuvel, J. C., and Cohen, A.,** pH influence on acidogenic dessimulation of glucose in an anaerobic digester, *Water Res.,* 16, 303, 1982.

55. **Massey, M. L. and Pohland, F. G.,** Phase separation of anaerobic stabilization by kinetic control, *J. Water Pollut. Control Fed.,* 50, 2204, 1978.

56. **Chynoweth, D. P.,** Microbial Conversion of Biomass to Methane, presented at the 8th Annual Energy Technology Conference and Exposition, Washington, D.C., 1981.

57. **Salkinoja-Salonen, M., Hakulien, R., Valo, R., and Apajalahty, J.,** Biodegradation of recalcitrant organochlorine compounds in fixed film reactors, *Water Sci. Technol.,* 15, 309, 1983.

58. **Cohen, A., Breune, A. M., Van Andel, J. G., and Van Deunsen, A.,** Influence of phase separation on the anaerobic digestion of glucose — 1) Maximum COD turnover rate during continuous operation, *Water Res.,* 14, 1439, 1980.

59. **Ghosh, S.,** Kinetics of Acid Phase Fermentation in Anaerobic Digestion in Proc. Third. Symp. on Biotechnology in Energy Production and Conservation, Scott, C. D., Ed., *Biotechnol. Bioeng. Symp. No. 11,* 301, 1981.

60. **Datta, R.,** Production of acid esters from biomass — novel processes and concepts, *Biotechnol. Bioeng. Symp.*, 11, 521, 1981.
61. **Levy, P. E., Sanderson, J. E., and Wise, D. L.,** Development of a process for production of liquid fuels from biomass, *Biotechnol. Bioeng. Symp.*, 11, 239, 1981.
62. **Rijkens, B. A.,** A novel two step process for the anaerobic digestion of solid wastes, in *Energy Biomass and Waste,* Vol. 5, Klass, D. L., Ed., Chicago, 1981, 463.
63. **Ghosh, S.,** Novel processes for high efficiency biodigestion of particulate feeds, Proc. Int. Conf. State-of-the Art on Biogas Technol. Transfer Diffusion, Cairo, November, 1984.
64. **Van den Berg, L., Kennedy, K. J., and Hamoda, M. F.,** Effect of type of waste on performance anaerobic fixed film and upflow sludge bed reactors, presented at 36th Industrial Waste Conference at Purdue University, Lafayette, 1981.
65. **Dahab, M. F. and Young, J. C.,** Energy recovery from alcohol stillage using anaerobic filters, *Biotechnol. Bioeng. Symp.*, 11, 381, 1981.
66. **Kennedy, K. J. and Van den Berg, L.,** Effects of temperature and overloading on the performance of anaerobic fixed film reactors, Proc. of 36th Int. Waste Conference, Purdue University, Lafayette, 1981.
67. **Kennedy, K. J. and Van den Berg., L.,** Continuous vs slug loading of downflow stationary fixed film reaction digesting piggery waste, *Biotechnol. Lett.*, 4(2), 137, 1982.
68. **Kennedy, K. J. and Van den Berg., L.,** Stability and performance of anaerobic fixed film reactors during hydraulic overloading at 10—35°C, *Water Res.*, 16, 1391, 1982.
69. **Van den Berg, L. and Kennedy, K. J.,** Effect of substrate composition on methane production rates of down flow stationary fixed film reactors, Proc. of VI Symp. Energy from Biomass and Waste, Florida, 1982.
70. **Binot, R. A., Bol, T., Noveau, H. P., and Nyns, E. J.,** Biomethanation by immobilized fluidized cells, *Water Sci. Technol.*, 15(8/9), 103, 1983.
71. **Shieh, W. K.,** Suggested kinetics model for fluidized biofilm reactor, *Biotechnol. Bioeng.*, 22, 677, 1980.
72. **Shieh, W. K. and Mulcahy, L. T.,** FBBR Kinetics — a rational design and optimization approach, *Water Sci. Technol.*, 15(8/9), 321, 1983.
73. **Hickey, S.,** Kinetics of acid-phase fermentation in anaerobic digestion, *Biotechnol. Bioeng. Symp.*, 11, 301, 1981.
74. **Frostell, B.,** Anaerobic fluidized bed experimentation with molasses wastewater, *Process, Biochem.*, 11, 37, 1982.
75. **Li, A. and Sutton, P. M.,** Dorr Oliver's anitron system — fluidized bed technology for methane production from daily wastes, presented at Whey Products Institute Annual Meeting, Chicago, 1981.
76. **Mosey, F. E.,** Methane fermentation of organic wastes, *Tribol. Cebedeau*, 453(34), 389, 1981.
77. **Jewell, W. J., Switzenbaum, M. S., and Morris, J. W.,** Minicipal wastewater treatment with the anaerobic attached microbial film expanded bed process, *J. Water Pollut. Control Fed.*, 53, 482, 1981.
78. **Morris, J. W. and Jewell, W. J.,** Organic particulate removal with the anaerobic attached film expanded bed process, presented at 36th Industrial Waste Conf. at Purdue University, Lafayette, 1981.
79. **Switzenbaum, M. S. and Jewell, W. J.,** Anaerobic attached film expanded bed reactor treatment, *J. Water Pollut. Control Fed.*, 52, 2953, 1980.
80. **Schraa, G.,** High rate conversions of soluble organic with the thermophilic anaerobic attached film expanded bed, presented at the 55th Annual Water Pollution Control Federation Conference, St. Louis, Missouri, 1982.
81. **Tait, S. J., Friedman, A. A.,** Anaerobic rotating biological contactor for carbonaceous wastewaters, *J. Water Pollut. Control Fed.*, 52(8), 2257, 1980.
82. **Edelmann, W.,** Technologies of developed countries and possibility of transferring them to developing countries, presented at International Conference State-of-the Art on Biogas Technology Transfer and Diffusion, Cairo, 1984.
83. **Martensson, L. and Frostell, B.,** Anaerobic wastewater treatment in a carrier assisted sludge bed reactor, *Water, Sci. Technol.*, 15(8/9), 233, 1983.
84. **Pfeffer, J. T.,** Methane fermentation engineering aspects of reactor design, *Tribol. Cebedeau*, 453(34), 357, 1981.
85. **Lettinga, G., Hobma, S. W., Pol, L. W. H., de Zeeuw, W. J., Grin, P., and Roersma, R.,** Design, operation and economy of anaerobic treatment, *Water Sci. Technol.*, 15(8/9), 177, 1983.
86. **Van den Berg, L. and Kennedy, K. J.,** Comparison of advanced anaerobic reactors, in Proc. Third. Int. Symp. on Anaerobic Digestion, Boston, 1983.
87. **Lettinga, G., Roersma, R., and Grin, P.,** Direct application of anaerobic treatment to sewage, H_2O, 14(10), 214, 1981.
88. **Hall, E. R.,** Biomass retention and mixing characteristics in fixed film and suspended growth anaerobic reactors, Proc. in IAWPR Specialized Seminar — Anaerobic Treatment of Wastewater in Fixed Film Reactors, Copenhagen, 1982.

89. **Norrman, J.,** Treatment of black liquor condensate form a pulp mill in an anaerobic filter and an expanded bed, *Water Sci. Technol.,* 15(8/9), 247, 1983.

90. **Young, J. C. and Dahab, M. F.,** Effect of media design on the performance of fixed bed anaerobic reactors, *Water Sci. Technol.,* 15(8/9), 369, 1983.

91. **Meunier, A. D. and Williamson, K. J.,** Packed bed biofilm reactors: simplified model, *J. Environ. Eng. Div.,* 107 (EE$_2$), 307, 1981.

92. **Feilden, N. E. H.,** The theory and practice of anaerobic digestion reactor design, *Process Biochem.,* 10, 34, 1983.

93. **Grin, P., Sutton, P. M., Henze, M., Elmaleh, S., Grady, C. P., Grujer, W., and Koller, J.,** Recomended notation for use in the description of biological wastewater treatment processes, *Water Res.,* 16, 1501, 1982.

94. **Hobson, P. N. Bousfield, S., and Summers, R.,** *Methane production from agricultural and domestic wastes,* Porteous, A., Ed., Applied Science London, 1981.

95. **Salkinoja-Salonen, M. S., Nyns, E. J., Sutton, P. M., Van den Berg, L., and Wheatley, A. D.,** Starting-up of an anerobic fixed film reactor, *Water Sci. Technol.,* 15(8/9), 3051, 1983.

96. **Van den Berg, L. and Lentz, C. P.,** Food processing waste treatment by anaerobic digestion, in Proc. of 32th Industrial Waste Conference at Purdue University, Lafayette, 1978.

97. **Goodrich, P.,** Personal communication, 1984.

98. **Benjamin, M. M., Ferguson, J. F., and Buggins, M. E.,** Treatment of sulfide evaporator condensate with an anerobic reactor, in *TAPP Environmental Conference,* New Orleans, 1981, 1.

99. **Elmaleh, S.,** Epuration et energie, *Tribol. Cebedeau,* 34, 83, 1981.

100. **Nyns, E. J., Naveau, H. P., and Demuynck, M.,** Biogas plants in the European Community and Switzerland: status, bottlenecks, future prospects, Proc. Third. Int. Symp. Anaerobic Digestion, Boston, 1983.

101. **Chen, S. C.,** Optimization of batch alcoholic fermentation of glucose symp. substrate, *Biotechnol. Bioeng.,* 23, 1827, 1981.

102. **Williams, D. and Munnecke, D. M.,** The production of ethanol by immobilized yeast cells, *Biotechnol. Bioeng.,* 123, 18(3), 1981.

103. **Scott, C. D.,** Ethanol production in a fluidized bed reactor utilizing flocculating *Zymonomas mobilis* with biomass recycle, *Biotechnol. Bioeng. Symp.,* 13, 287, 1983.

104. **Gomma, G., Uribellardea, J. L., Sovcaille, P., Minier, M., and Ferras, E.,** High cell concentration culture: physiological factors affecting this parameter and technological solution, 3rd European Congress on Biotechnology, Munchen, 1984.

105. **André, G., Moog-Young, M., and Robinson, C. W.,** Improved method for the dynamic measurement of mass transfer coefficient for application to solid-substrate fermentation, *Biotechnol. Bioeng.,* 123, 1611, 1981.

106. **Jewell, W. J., Chandler, J. A., Dell'Orto, S., Fanfoni, K. J., Fast, S. J., Jackson, D., and Kabrick, R. M.,** Dry fermentation of agricultural residues, SERI/XB-0-9038-H6, April 1981.

107. **Leuschner, A. P. and Melden, H. A.,** Landfill enhancement for improving methane production and leachate quality, Proc. 56th Annual Conference of Water Pollution Control Federation, Atlanta, 1983.

108. **Wise, D. L., Leuschner, A. P., Levy, P. F., and Sharaf, M. A.,** A large-scale biologically derived methane process, presented at M.I.T. Summer Course Biotechnology 20.405, Massachusetts, 1982.

109. **Reese, E. T. and Mandels, M.,** Stability of the cellulase of *Trichoderma Reesei* under use conditions, *Biotechnol. Bioeng.,* 22, 323, 1980.

110. **Tsao, G. T.,** Pretreatment of cellulosic material for saccharification, microbiology applied to biotechnology, Proc. 12th International Congress of Microbiology, Münich, 1979.

111. **Datta, R.,** Acidogenic fermentation of corn stover, *Biotechnol. Bioeng.,* 123, 61, 1981.

112. **Wentworth, R. P., Wise, D. L., and Levy, P. F.,** Production of liquid fuels and chemicals through anaerobic digestion of biomass. I. *Chem. E. Symp. Series,* 78, 1982.

113. **Datta, R.,** Acidogenic fermentation of lignocellulose, *Biotechnol. Bioeng.,* 23, 2167, 1981.

114. **Omstead, D. R., Jeffries, T. W., Naugton, R., and Gregor, H. P.,** Membrane controlled digestion: anaerobic production of methane and organic acids, *Biotechnol. Bioeng. Symp.,* 10, 247, 1980.

115. **Jewell, W.,** Personal communication, 1984.

Chapter 4

BIOGAS FROM SANITARY LANDFILL TECHNIQUE IN EGYPT

M. N. Alaa El-Din, E. A. Saleh, Y. Z. Ishac, S. A. El-Shimi, and M. El-Housseini

TABLE OF CONTENTS

I.   Abstract 132

II.   Introduction 132

III.   Materials 132
      A.   Garbage 132
      B.   Sewage Sludge 133

IV.   Experimental Procedure 133

V.   Results and Discussion 133

References 143

I. ABSTRACT

Cairo, with 8 million inhabitants generates about 4300 tons of garbage per day. Dumping and burning of this garbage not only causes air pollution but also the loss of badly needed organic manure. Sewage sludge accumulating from treatment plants is dried using inefficient and hazardous methods.

The present study is aimed at evaluating the sanitary landfill concept as a source for biogas and compost, and as a tool for improving health and the environment.

Different laboratory experiments were conducted to ferment city garbage in combination with water or with sewage sludge at different levels — 10, 25, and 35% TS (total solids). Fermentation took place in 1.25 and 3.5 ℓ fermentors. Some of the fermenters were kept at 35 and 55°C. The experiments lasted for 4 to 6 months. Gas volume, methane content, pH, soluble-N, and contents of total and volatile solids were estimated regularly. Counts of anaerobic acid formers and cellulose decomposers as well as the survival of coliform and salmonella groups of pathogenic bacteria were also estimated. The data were discussed in view of possible solutions for garbage and sewage problems of large cities in Egypt.

II. INTRODUCTION

The Cairo area generates about 4300 ton/day (tpd) municipal solid waste (MSW), about 1500 tpd are collected and recycled by private garbage collectors for feeding pigs and preparing organic manure at very severe conditions. The remaining 2800 tpd are collected by governmental systems to be burned in several dumping places thus losing energy, fertilizers, and soil conditioners as well. Air pollution is also one of the threats. Alaa El-Din et al., estimated that sanitary landfills combined with production of high Btu gas and soil conditioners could be a good solution for the increasing MSW accumulated in Cairo and other cities of Egypt.[1]

However, their figures were of theoretical nature. Biogas and soil conditioners from sanitary landfills of MSW were found to be the cheapest option as compared with the composting and burning of it.

At present it is estimated that 80% of the urban Cairo is sewered, with 65% of the population connected. There is no sufficient capacity to handle the wastewater now collected. It is also assumed that 472 tons of volatile solids per day could be made available for energy production through anaerobic digestion. Expected energy produced was calculated[2] to be about 1.4×10^{12} J/day. However, the cost of plants is high.

The combined fermentation of MSW and sewage sludge for production of biogas and soil conditioners became an attractive option as Alaa El-Din et al., found improved biogas production from mixtures of garbage and sewage sludge in different Chinese and Indian biogas digesters.

The present study is aimed at evaluating the biogas productivity, the production of soil conditioners, the behavior of different types of bacteria with special reference to pathogenic ones and the changes occurring during the fermentation of MSW in a sanitary landfill system.

III. MATERIALS

A. Garbage

MSW was collected from three big vegetable markets of Cairo, dried and ground. The chemical composition of the ground garbage was as follows: moisture content, 10.020%; volatile solids, 78.270%; organic carbon, 45.397%; total nitrogen content, 0.945%; and C/N ratio, 49.61.

B. Sewage Sludge

Sewage sludge was collected from the biggest sewage treatment plant of Cairo at Zenin. The chemical analysis was as follows: total solids, 2.269%; volatile solids, 68.089%; organic carbon, 39.492%; total nitrogen, 1.981%; and C/N ratio, 19.935.

IV. EXPERIMENTAL PROCEDURE

Three sets of experiments were prepared to evaluate the behavior of different concentrations of garbage, sewage sludge, and water when fermented at room temperature (10 to 30°C), at 35°C (mesophilic), and at 55°C (thermophilic). The garbage was mixed with either water or sewage sludge to satisfy three concentrations of total solids, namely, 10, 25, and 35%. Fine pulverized $CaCO_3$ was mixed thoroughly with the garbage to satisfy 10% of the TS in each fermenter. The mixtures fermented at room temperature were poured in 1.25 ℓ capacity fermenters, while those incubated at 35°C were put in 3.5 ℓ fermenters and kept in a walk-in incubator (10 m³ capacity) and the thermophilic incubated fermenters were kept in three similar incubators adjusted at 55°C. The amount of the fermenting material differed according to fermentor size, giving the same size proportion of fermenting material to fermenter. Six replicable fermenters were prepared for each treatment, totaling 108 fermenters running at the same time.

After weeks 0, 1, 2, 4, and 14 of incubation, the content of one of the fermenters was thoroughly mixed and representative samples were analyzed for the following.

1. pH; in 1:2 water extract using glass electrode pH-meter.
2. Ammoniacal-N, in the extract made by one normal solution of KCL by distillation with MgO according to Piper.[14]
3. Contents of total and volatile solids according to the standard methods.[3]
4. Counts and survival of the following groups of microorganisms in the fermenting material. (A) MPN of acid formers on nutrient broth enriched with 5 g glucose per liter and bromothymol blue solution as indicator and Durhams tubes.[5] (B) MPN of cellulose decomposers.[4,13] (C) Counts of *E. coli* by plate counting on MacConkey's medium.[6] (D) Counts of *Salmonella* and *Shigella* by plate counting on SS agar medium.[6]

Gas yield was measured every day for up to 23 weeks of fermentation. Gas samples were analyzed for their methane and carbon dioxide contents every week using the following methods.

Methane content — Methane content was estimated weekly using the following technique as described by Walter and Williams.[18] Gas samples were withdrawn into a 50 mℓ syringe and 0.5 mℓ gas samples were injected into GOW MAC gas chromatograph model 750P, fitted with a 120 cm long and 0.2 cm diameter stainless steel column filled with 5% OV-101 on CHROM-PAW 80-100 mesh and with dual flame ionization detector. The carrier gas was nitrogen at a flow rate of 28 mℓ/min. Hydrogen was generated by an attached hydrogen generator and provided at the rate of 30 mℓ/min. Air at the rate of 300 mℓ/min was applied for the flame. The operation temperatures were 75°C for column oven, 100°C for injection port, and 150°C for detector. Standard curves were prepared using pure methane gas. (Messer Griessheim GMBH, Frankfurt, FRG) and were used as a reference for calculating methane concentration in the biogas produced.

Carbon dioxide — Co_2 content was measured weekly by Orsat's apparatus using caustic potash solution for CO_2 absorption.[3]

V. RESULTS AND DISCUSSION

The average daily production of biogas from garbage presented as m³ biogas per ton of

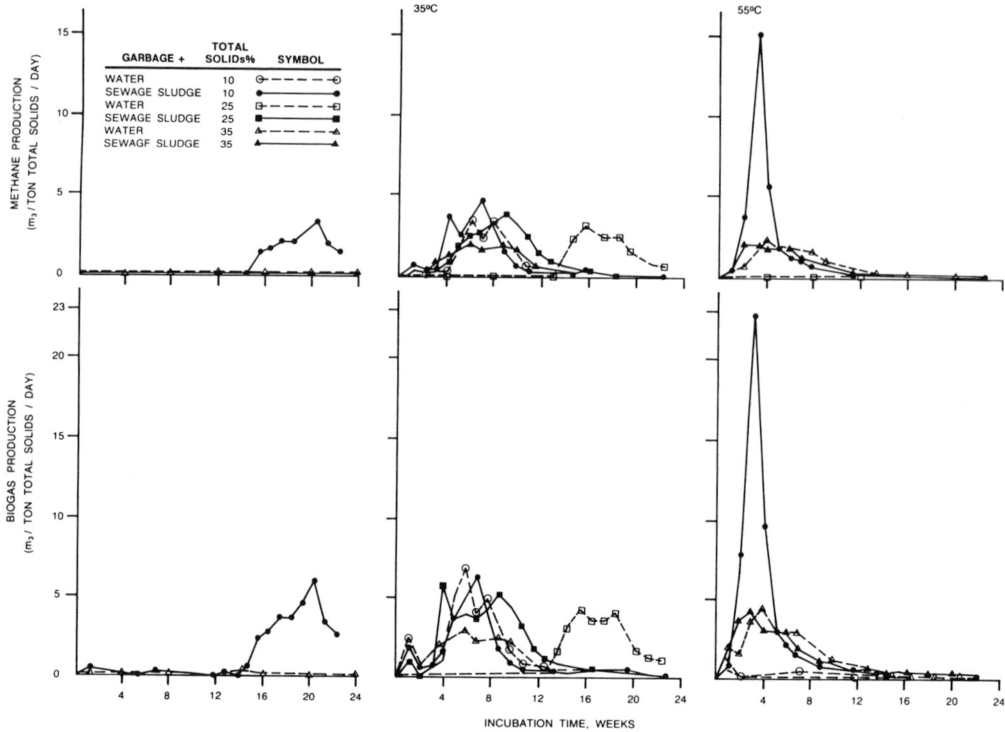

FIGURE 1.   Daily production rate of biogas and methane (m³/ton TS) from sanitary landfill system containing different amounts of garbage fermented in combination with water or sewage sludge at different temperatures.

total solids per day (m³/ton/day) and the average daily yield of methane (m³/ton/day) are illustrated in Figure 1. The data showed that at room temeprature no considerable gas production was recorded for any of the garbage concentrations, except for 10% garbage combined with sewage sludge. The production started after 14 weeks of incubation and reached its maximum after 21 weeks, being about 6 m³/ton/day. The maximum methane production rate was also reached after the same period of incubation being about 3 m³/ton/ day. The gas generation and the yield of methane as well decreased rapidly thereafter. The incubation of garbage mixtures at 35°C led to an early production of biogas and methane, except in the case of high concentration of garbage moistened with sewage sludge. Garbage at 35% TS moistened with water showed delayed gas production — activity started after 23 weeks of incubation. The gas produced during the first 2 weeks was very poor in methane content. However, starting from the 5th week, the methane yield increased as a result of increasing both gas yield and methane content. The generation of biogas decreased by all treatments rapidly after reaching the maximal values. The thermophilic biogas generation seemed to favor the low concentration of garbage (10% TS) receiving sewage sludge for moistening. The highest values for biogas produced (about 23 m³/ton/day) and for methane generated (about 15 m³/ton/day) were recorded after 3 weeks of incubation. However, the gas and methane production decreased very rapidly. Higher concentration of garbage either did not show any biogas generation (25% TS) or produced much less biogas and methane. The peak of production was recorded during the same period of fermentation as for the lower concentration.

The accumulated gas production of fermented garbage showed more clear and differen-tiated figures (Figure 2). The total biogas production was (after 23 weeks) almost the same for garbage mixed with sewage sludge (10% TS) when incubated at room temperature or

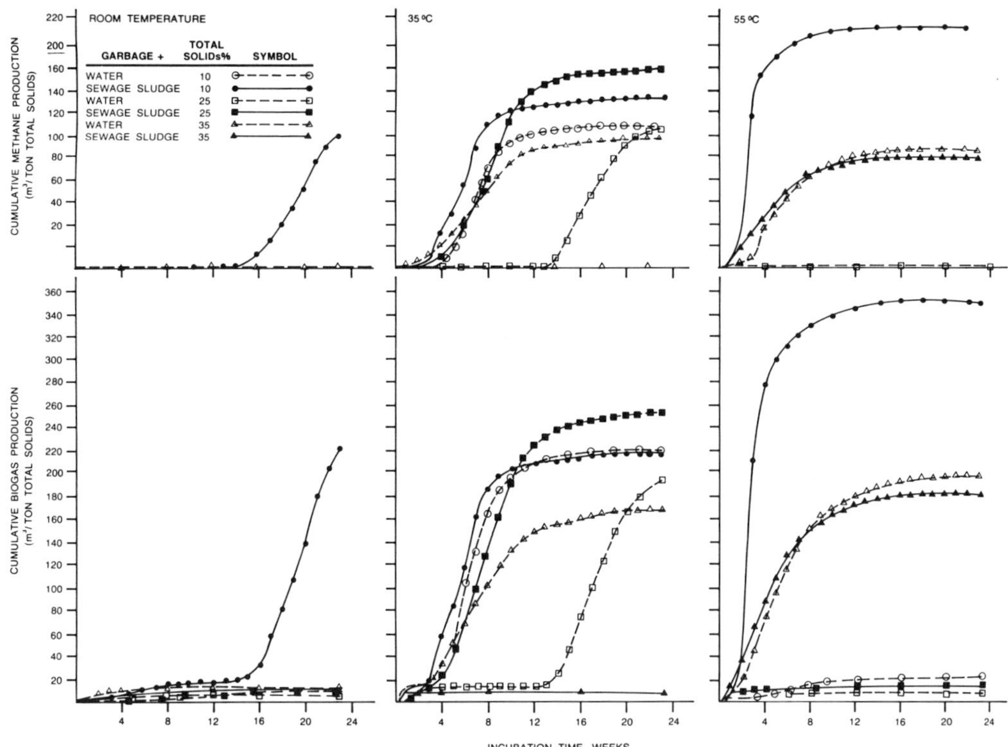

FIGURE 2.   Cumulative biogas and methane production (m³/ton TS) from fermentation of different loading rates of garbage combined with water or sewage sludge in sanitary landfill system at different temperatures.

at 35°C, being 220 m³/ton solids/23 weeks. The difference was in the time needed to start active biogas generation. This was about 3 and 15 weeks when incubated at 35°C and room temperature respectively. The total methane yield was higher for the same garbage concentration when fermented at 35°C (about 150 m³/ton/23 weeks) than incubated at room temperature (110 m³/ton/23 weeks). Similar findings were reached by Naser[12] using cow dung. She stated that the temperature affected the biogas production and the detention time but not the final volume of biogas produced or the amounts of volatile solids destroyed. Under thermophilic condition (55°C) Velsen et al.[16] found that methane production decreased by about 25% as compared to the mesophilic digestion when using pig manure. In the present study the opposite was recorded for the low concentration of garbage (10% TS). The difference could be attributed to the substrate used in those experiments. Wise et al.[19] reported that an increase of biogas production by 10% from the sanitary landfill when the temperature increased from mesophilic to reach thermophilic temperature (60°C).

Application of sewage sludge to garbage gave higher yields of biogas. The differences were much larger when considering the methane concentration. This was true for all temperatures and especially for low solid concentrations. High garbage concentration (35%) did not show this difference when incubated at 35°C.

The changes in methane contents in biogas produced during the fermentation of garbage mixed with water or sewage sludge and incubated at different temperatures are illustrated in Figure 3. The data showed in general that fermentation of garbage at room temperature generated a low concentration of methane in the biogas. The lowest values were recorded for garbage moistened with water. This indicates the stimulation effect of sewage sludge which acted as a good starter for methane fermentation. The best starter effect of sewage sludge on methane concentration was recorded for the lowest loads of solids (10%) and

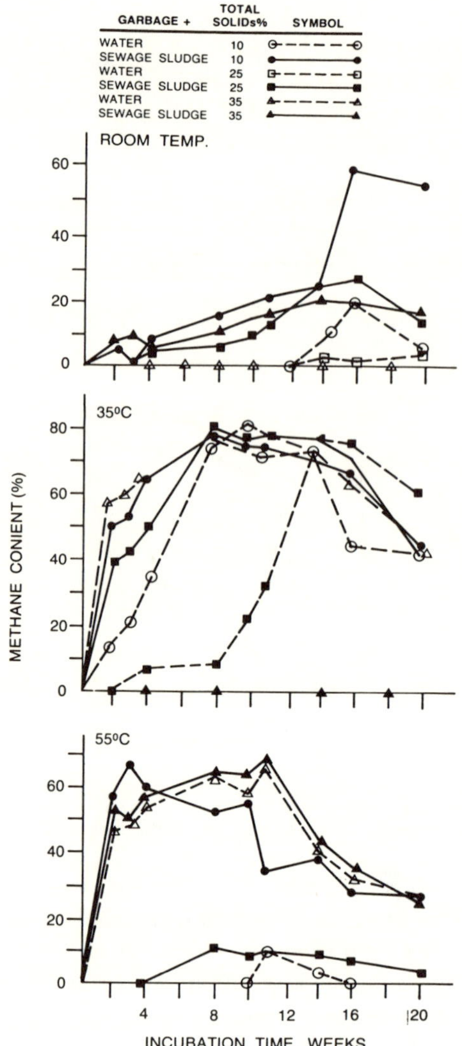

FIGURE 3.   Changes in methane content of biogas produced from garbage in combination with water or sewage sludge incubated at different temperatures in sanitary
landfill system.

about 61% methane was estimated after 16 weeks, while the higher loads of solids gave
only 28 and 20% methane for the concentration 25 and 35%, respectively. Hobson and
Show,[8] Velsen,[16] Wise et al.,[16] Alaa El-Din[2] also reported the stimulative effect of digested
sludge as a good starter and Wise attributed this to its role in supplying the system with
high numbers of acid formers and methanogenic bacteria, nitrogen, phosphorus, and buffering agents.

Increasing the incubation temperature to reach the mesophilic optimal range (35°C) did
help to increase the methane concentration in the biogas produced. This was also reached
after shorter periods. Sewage sludge also proved to increase the methane concentration at
faster rates.

The thermophilic methane formation was very low in water treated garbage at low or
medium loading rates (10, 25 and 35% TS). Sewage sludge application improved the situation

Table 1
CHANGES IN pH DURING THE FERMENTATION OF DIFFERENT LOADING RATES OF GARBAGE COMBINED WITH WATER OR WITH SEWAGE SLUDGE IN A SANITARY LANDFILL SYSTEM INCUBATED AT DIFFERENT TEMPERATURES

Sampling time (weeks)	Total solids (%) source of moisture					
	10		25		35	
	Water	Sludge	Water	Sludge	Water	Sludge
Initial	7.15—7.30	6.4—6.45	8.0—8.35	7.3—7.95	8.25—8.8	7.1—7.6
Incubation at room temperature						
1	6.10	6.00	6.45	6.10	6.85	6.95
2	5.80	5.50	6.20	5.90	6.20	6.80
4	5.90	5.70	6.20	6.30	6.90	6.90
14	5.20	5.10	5.60	5.70	6.00	6.10
Incubation at 35°C (mesophilic)						
1	5.80	5.60	7.30	6.85	7.10	7.10
2	5.40	5.40	6.80	6.90	6.90	7.30
4	5.40	5.40	7.00	6.90	8.70	6.50
14	7.90	8.15	6.45	7.75	8.55	6.70
Incubation at 55°C (thermophilic)						
1	6.20	6.40	6.30	7.10	7.30	7.35
2	7.2	8.20	7.70	7.00	7.50	7.20
4	6.1	8.80	7.00	6.90	8.00	8.20
14	5.8	8.45	6.30	6.30	8.20	8.40

especially at low and high concentrations of garbage (10 and 35%). Water treated garbage showed surprisingly high methane content in the biogas produced as compared with other water treatments.

In general, the pH values presented in Table 1 showed a sharp increase in the hydrogen ion concentration during the first week of incubation especially in the fermenters treated with water while those receiving sewage sludge showed moderate increase in hydrogen ion concentrations or moderate decrease in pH values. This occured although $CaCO_3$ was added at the rate of 10% of the total solids. There was, however, no clear relation between pH values and either the biogas generation rate or the concentration of methane. Many of the fermenters showed a pH range favorable to biogas generation and methane formation although little or no biogas production could be recorded. The analysis of fatty acids carried out so far showed accumulation of high concentrations of propionic and butyric acids in those fermenters.

Hobson et al.[9] made propionic acid — the principal component of the volatile fatty acids (VFA) — responsible for the failure of production or for low productivity of biogas. Accumulation of propionic acid and the increase in VFA concentration was found by Velsen[15] to be as a result of increasing the load of volatile solids. Accumulation of ammnonia in the digesting material beyond the concentration of 1500 ppm was found by Fisher et al.[7] and Lapp et al.[10] to be responsible for increased accumulation of fatty acids.

In present experiments the concentrations of ammonia (Figure 4) were in some cases much higher than 1500 ppm. Application of sewage sludge to the garbage increased the

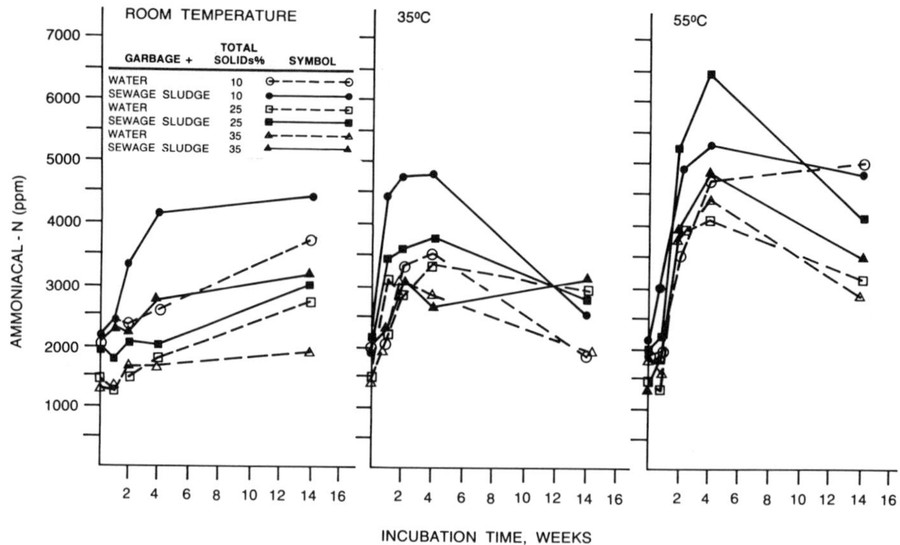

FIGURE 4.   Changes in ammoniacal *N* during the fermentation of different loading rates of garbage combined with water or sewage in a sanitary landfill system at different temperatures.

ammonia concentration in the fermenting material. General ammonia concentration increased by increasing the fermentation time to reach a maximal value after 14 weeks of incubation at room temperature, while this value was reached after 4 weeks when incubated at 35 and 55°C. The concentration of ammonia decreased thereafter. The concentration of ammonia was after 14 weeks of incubation still higher than the initial one.

Ammonia concentrations higher than 3000 ppm were reported to be inhibitory to generation of biogas without affecting the production of VFA.[11] The digesting material was diluted pig manure. In the present study, a higher solids concentration of garbage was fermented and no clear relation was found between increasing ammonia concentration to reach more than 6000 ppm and the biogas generation. The fact is that in the presence of a sufficiently high pH (above 7.5), the equilibrium between ammonia form (NH_3) and ammonium ion (NH_4^+) is shifted to the side of ammonia predominance, which in turn is less inhibitory to biogas generation than ammonium ion.[20] This fact could not help explain the differences in gas production in the present study based on ammonia-ammonium balance.

The relation between biogas generation and the disappearance of volatile solids is given in Table 2. The percent of volatile solids (VS) destroyed showed clearly the optimal effect of the mesophilic temperature range for incubation on both gas and methane generation. The stimulative effect of sewage sludge on gas yield of fermenting garbage was also proven. The data showed that the rate of biogas production from VS added was in the sanitary landfill system close to those recorded for typical biogas digesters. The rates recorded in the present study ranged between 217 and 377 ℓ biogas/kg VS added. Naser[12] recorded a sharp decline from 349 to 102 ℓ biogas/kg VS when the TS content was increased from 8.2 to 18.3% and incubated at 37°C in regular fermenters. In the present study 25 and 35% solids gave 225 to 312 ℓ biogas/kg VS. The duration, however, was much longer in the latter to match the sanitary landfill. Similar digestion rates were reported by Wise et al.[19] using shared MSW and sewage sludge, which indicated a 0.1 to 30% destroyment of VS under different incubation conditions (37 to 60°C). In some fermenters up to 72.1% of VS were converted to biogas.

The average daily production rates of biogas and methane referring to the volume of the digesting material were also calculated to provide some comparison with digesting systems

Table 2
BIOGAS AND METHANE PRODUCTION IN RELATION TO THE CONCENTRATION OF SOLIDS DURING FERMENTATION OF GARBAGE IN SANITARY LANDFILL SYSTEM FOR 23 WEEKS AT 3 DIFFERENT TEMPERATURES

Treatments		Volatile Solids			Biogas produced ℓ/fermenter	Methane produced ℓ/fermenter	Biogas ℓ/kg V.S. added	Methane ℓ/kg V.S. added
% of Solids as garbage	Source of moisture	Initial	Final	% V.S. destroyed				
Incubation at room temperature								
10	Water	38.744	38.380	0.940	0.439	0.020	11.331	0.516
10	Sludge	46.310	37.169	19.739	11.045	5.805	238.501	125.351
25	Water	105.664	105.000	0.628	0.803	0.001	7.600	0.012
25	Sludge	110.521	109.333	1.075	1.435	0.146	12.984	1.321
35	Water	144.408	142.439	1.363	2.379	0	16.474	0
35	Sludge	147.333	145.662	1.134	2.018	0.174	13.697	1.181
Incubation at 35°C (mesophilic)								
10	Water	154.975	118.723	23.392	43.804	25.094	282.652	161.923
10	Sludge	181.784	145.605	19.902	43.716	29.622	240.483	162.95
25	Water	464.924	369.902	20.427	114.759	74.023	246.834	159.215
25	Sludge	482.978	358.285	25.818	150.670	103.939	311.960	215.704
35	Water	563.544	464.516	17.927	122.075	82.509	216.620	146.411
35	Sludge	575.631	569.988	0.980	6.819	0	11.846	0
Incubation at 55°C (thermophilic)								
10	Water	38.744	37.891	2.203	1.032	0.009	26.636	0.232
10	Sludge	46.310	31.865	31.192	17.454	10.613	376.895	229.173
25	Water	105.664	104.812	0.807	1.031	0.019	9.959	0.180
25	Sludge	110.521	109.138	1.251	1.671	0.013	15.121	0.118
35	Water	144.408	114.444	20.750	36.200	19.329	250.679	133.850
35	Sludge	147.333	111.931	18.599	33.111	18.196	224.736	123.503

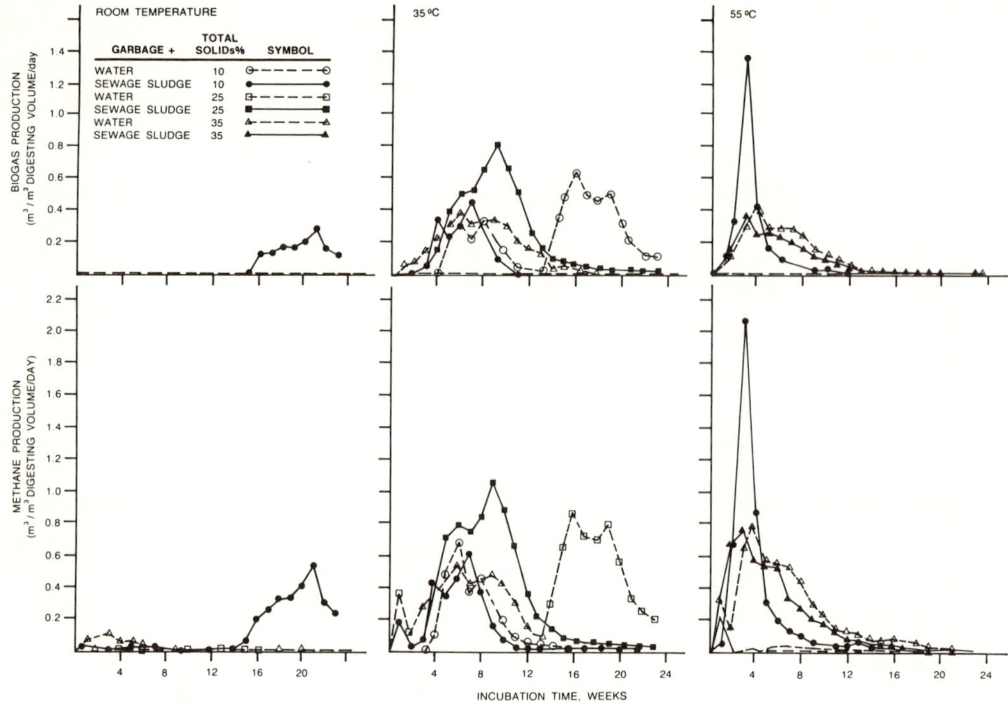

FIGURE 5.   Daily production rate of biogas and methane (m³/m³ digesting volume/day) from sanitary landfill system containing different amounts of garbage fermented in combination with water or sewage sludge at different temperatures.

used for biogas generation. The data are presented in Figures 5 and 6. The data showed a similar trend for the production rates based on the TS added. The productivity of unit digesting volume of the landfill increased with time, reaching their maximum after 21 weeks of incubation at room temperature (0.55 v/v/day for biogas and 0.3 v/v/day for methane), or after 6 to 16 weeks of incubation at 35°C (0.55 to 1.8 v/v/day for biogas, 0.53 to 0.85 v/v/day for methane). Thermophilic incubation gave early high rates (4 weeks incubation) ranging between 0.75 to 2.1 v/v/day for biogas and from 0.35 to 1.4 v/v/day for methane.

The cumulative figures were calculated to provide a clear idea about the rate of production per unit volume of the landfill throughout the 23 week period of incubation (Figure 6). This type of calculation can help in calculating the expected gas yield from a known volume of sanitary landfill in a certain period of time.

Wise et al.,[19] recorded also the highest gas production rate (about 0.82 v/v/day methane) after 30 days of incubation at 37°C and 25% solids. Thereafter, the production decreased sharply to reach about 0.2 v/v/day after 90 days of incubation. This was still higher than methane production rates recorded for optimal operation of indian type biogas digesters and more than double the rates recorded for chinese type digesters.[2]

The acid producing bacteria (Figure 7) increased in numbers to reach their maximum after 1, 2, or 4 weeks of incubation at room temperature. Thereafter, the numbers decreased rapidly to reach numbers which were still higher than at the start. Incubation at higher temperatures (35 and 55°C) increased the numbers slightly. In general the application of sewage sludge favored the growth of acid-forming bacteria, as the sludge provided the system with both inoculum and nutrients needed. Increasing the load of solids was also accompanied with higher numbers of acid formers. In the thermophilic range, high concentrations of garbage showed lower numbers of acid formers at the early stages of growth.

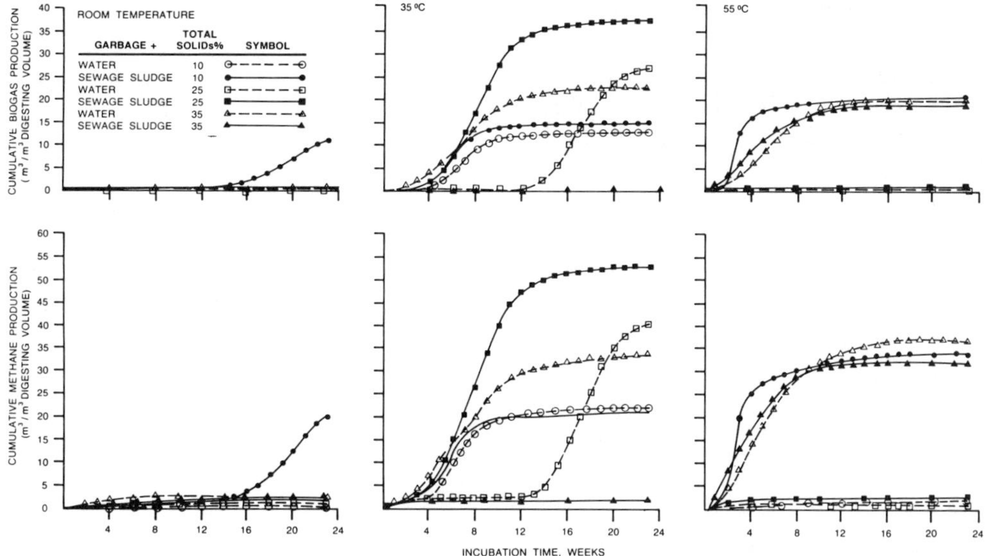

FIGURE 6.   Biogas and methane production (m³/m³ digesting volume) from the fermentation of different loading rates of garbage combined with water or sewage sludge in sanitary landfill system at different temperatures, cumulative curves.

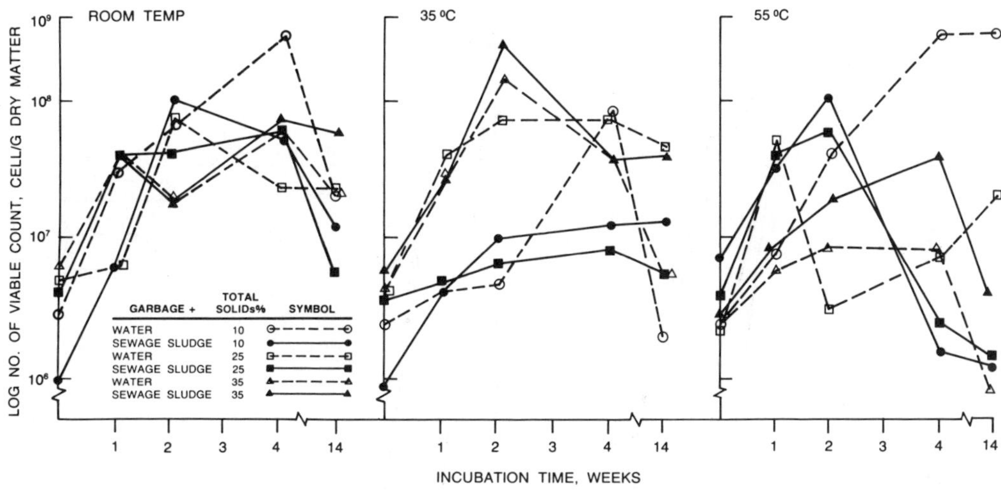

FIGURE 7.   Changes in populations of acid producing bacteria during fermentation of garbage moistened with water or sewage sludge at different temperatures.

There was, however, no clear relation between the numbers of acid-forming bacteria and the activity of biogas generation.

The anaerobic cellulose decomposers (Figure 8) showed almost the same trend as the acid formers. Their numbers were much lower than the acid formers, but they also did not show clear relation to biogas production.

The survival of phathogenic bacteria *Salmonella* and *Shigella* (Figure 9) showed a rapid reduction in their numbers indicating complete disappearance within 4 weeks of incubation

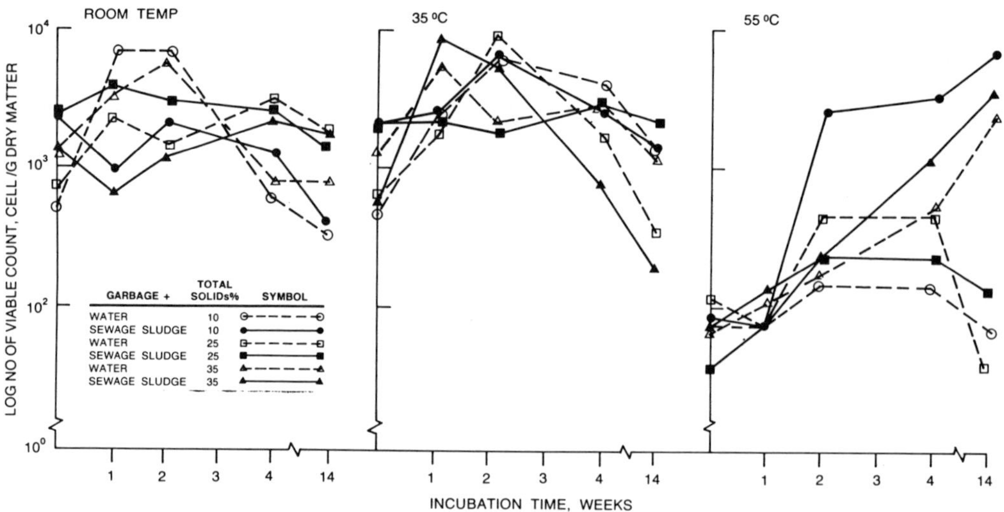

FIGURE 8.   Changes in populations of cellulose decomposers during fermentation of garbage moistened with water or sewage sludge at different temperatures.

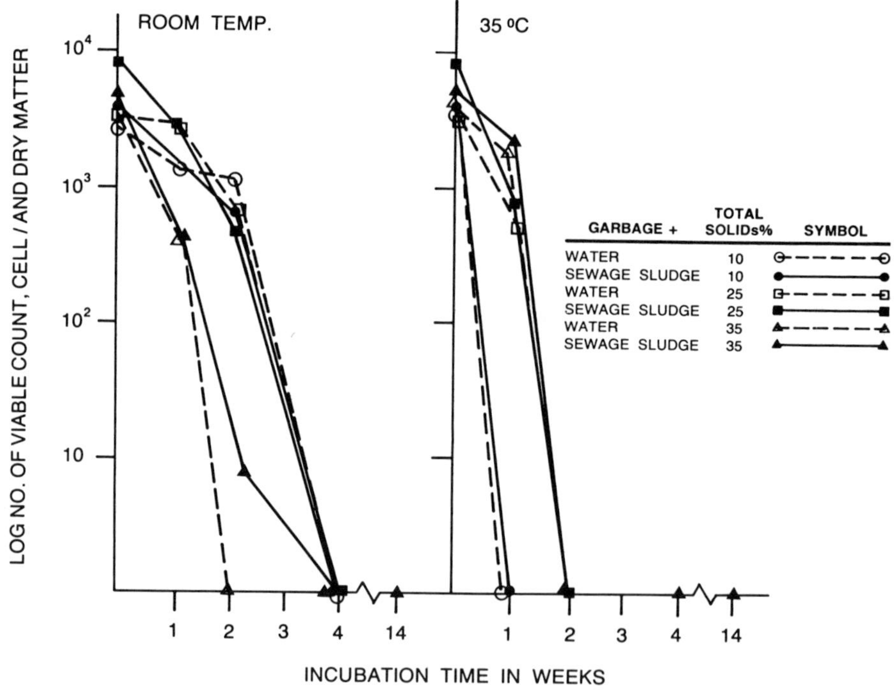

FIGURE 9.   Survival of pathogenic bacteria (*Salmonella* and *Shigella*) in garbage fermented in combination with water or sewage sludge in sanitary landfill system at different temperatures.

at room temperature (10 to 21°C). This was achieved at earlier time (2 weeks) when incubated at 35°C. The thermophilic incubated fermenters did not show any of the *Salmonella* or *Shigella* from the very beginning. Garbage moistened with sewage sludge showed higher loads of *Salmonella* and *Shigella*. Increasing the solids concentration provided same protection for *Salmonella* and *Shigella* when incubated at 35°C. They disappeared from the low solid loaded fermentor (10%) during the first week of incubation. Their numbers were higher

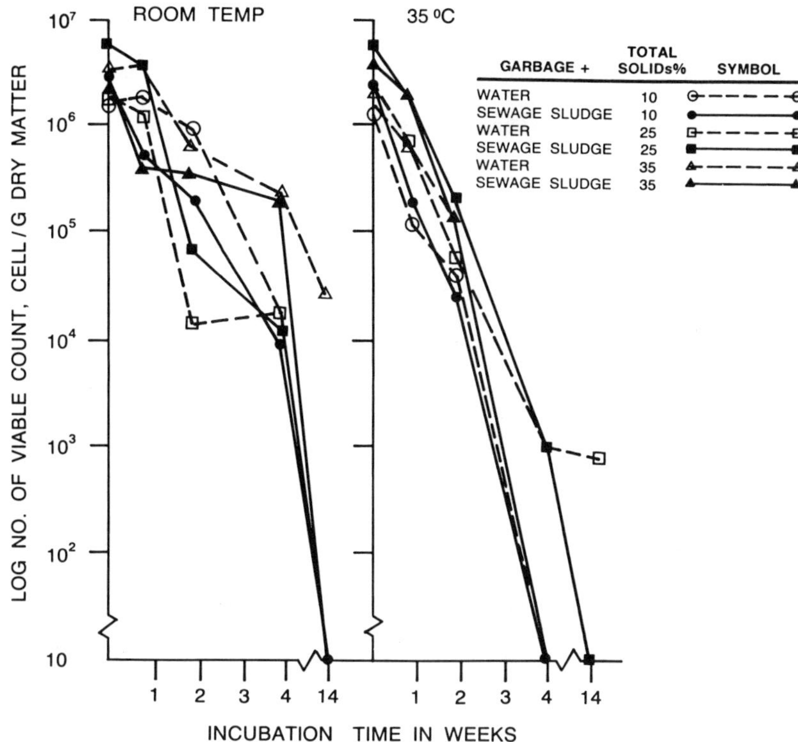

FIGURE 10.   Survival of pathogenic bacteria (coliform group) in garbage fermented in combination with water or sewage sludge in sanitary landfill system at different temperatures.

in the heavily loaded fermenters (35% TS) than in those loaded with 25% TS. Similar protective effect of high solids was recorded to pathogenic bacteria by Naser.[12] She reported that longer time was needed for the destroying of pathogenic bacteria when the concentration of solids in the digesting mixtures was increased. Similar results were obtained by El-Housseini[21] for sewage sludge.

The counts of *E. Coli* (Figure 10) decreased rapidly by increasing the time of incubation. Their numbers were, much higher than *Salmonella* and *Shigella*, where the rate of disappearance was lower than the other group, but it was higher when incubated at 35°C than at room temperature.

REFERENCES

1. **Alaa El-Din, M. N., Abu-Talib, and Fritz, J.,** Biomass energy potential in Egypt. A realistic Appraisal. Bioenergy, 80, World Congress, Atlanta, Georgia, April, 21 to 24, 1980.
2. **Alaa El-Din, M. N. Ishac, Y. Z., Saleh, S. A., El-Shimi, S. A., and El-Houssini, M.,** Biogas from city refuses in Egypt, The first OAU/STRC inter-African Conference on Bio-fertilizers, Cairo, Egypt, March 22 to 26, 1982.
3. American Public Health Association, *Standard Methods for the Examination of Water and Waste-water,* 14th ed., APHA, Washington, D.C., 1976.
4. **Allen, O. N.,** *Experiments in Soil Microbiology,* Burgess Publ., Minneapolis, Minn., 1959.
5. **Cunningham, A.,** *Practical Bacteriology,* 2nd ed., Oliver and Boyd, London, 1954.
6. Difco Manual and Dehydrated Culture Media and Reagents, 8th ed., Difco Laboratories, Detroit, Mich., 1977.

7. **Fisher, J. R, Lanotti, E. L., Proter, J. H., and Garcia, A.,** Producing methane gas from swine manure in a pilot — size digester, ASAE Paper No. MC 77 — 604, March, 1977.

8. **Hobson, P. N. and Show, B. G.,** The anaerobic digestion of waste from an intensive pig unit, *Water Res.,* 7, 437, 1973.

9. **Hobson, P. N., Bousfield, S., and Summers, R.,** Anaerobic digestion of organic matter, *CRC. Crit. Rev. Environ. Control.,* 6, 131, 1974.

10. **Lapp, H. M., Sparling, A. B., Schult, D. D., and Buchaman, L. C.,** Methane production from animal wastes. I. Fundamental considerations, *Can. Agric. Eng.,* 17, 97, 1975.

11. **Melbinger, N. R. and Donnullon, J.** Toxic effects of ammonia nitrogen in high-rate digestion, *J. Water Pollution Control Fed.,* 43 (8), 1658, 1971.

12. **Naser, A. F.,** Production of Methane from Biomas, Ph.D. theses, Faculty of Engineering, Cairo University, Egypt, 1980.

13. **Omeliansky, W. L.,** *Zentralbl. Bakteriol. Parasitenkd. Abt.,* 118, 224, 1902.

14. **Piper, C. S.,** *Soil and Plant Analysis,* Interscience, New York, 1950.

15. **Van Velsen, A. F. M,** Anaerobic digestion of piggery waste. I. The influence of detention time and manure concentration, *Neth. J. Orgic. Sci.,* 25, 151, 1977.

16. **Van Velsen, A. F. M.,** Anaerobic digestion of piggery waste. II. Start up procedure, *Neth. J. Agric. Sci.,* 27, 142, 1979.

17. **Van Velsen, A. F. M., Lettinga, G., and den Ottelander, D.,** Anaerobic digestion of piggery waste. III. Influence of temperature, *Neth. J. Agric. Sci.,* 27, 255, 1979.

18. **Walter, J. W. and Williams, J. J.,** Dry anaerobic fermentation, in *Biotechnol. Bioengin. Symp. No 10,* John Wiley & Sons, New York, 1980.

19. **Wise, D. L., Blamchet, M. J., Boyd, W. F., and Pacey, J. G.,** Fuel gas enhancement by controlled of municipal solid waste, *Res. Conserv.,* 6, 3, 1981.

20. **Wujcik, W. J. and Jewell, W. J.,** Dry anaerobic fermentation, *Bio. Symp. No. 10,* John Wiley & Sons, New York, 1980.

21. **El-Houseini, M.,** Unpublished data.

Chapter 5

THE U.A.S.B. — REACTOR TREATING PAPER — AND BOARD MILL EFFLUENT

P. J. F. M. Hack

TABLE OF CONTENTS

I.   Introduction 146

II.   Anaerobic vs. Aerobic Treatment 146

III.   Biochemistry 146

IV.   Treatment Methods 146

V.   The U.A.S.B.-Reactor 149

VI.   The U.A.S.B.-Reactor Treating Paper Mill Effluents 150
       A.   Introduction 150
       B.   Purpose..150
       C.   Results 150
              1.   Start-up 150
              2.   COD-Removal 152
              3.   Biogas 152
              4.   Volatily Fatty Acids 152

VII.   The Full-Scale Plant at Papierfabriek Roermond 152
       A.   Description of the Plant 152
       B.   Cost Analyses 152

VIII. Conclusions...153

References...154

I. INTRODUCTION

During the last decade great progress has been made on anaerobic wastewater treatment. Different reactortypes have been developed and compared to aerobic methods, the advantages are numerous. Until recently almost all applications of the U.A.S.B.-principle for anaerobic wastewater treatment could be found at food-processing industries. Only in 1983 were the first full scale U.A.S.B.-reactors started in the paper industry. In this paper research and operating results are presented.

II. ANAEROBIC VS. AEROBIC TREATMENT

The anaerobic method of waste treatment under the present circumstances, offers a number of significant advantages with few serious or insuperable drawbacks over other treatment methods. Conventional benefits and limitations of the process have been summarized in Table 1. At the present state of process technology, there are few drawbacks of the anaerobic process left, and all of its benefits are still valid, as will be shown in this paper, and others.[4]

III. BIOCHEMISTRY

Basically, anaerobic digestion of complex organic matter to methane consists of four successive steps (Figure 1) according to de Zeeuw.[3]

1. Hydrolysis — where organic polymers are hydrolized to their individual monomers as a result of enzymatic attack.
2. Acid formation — the hydrolyzed compounds are converted intracellulary by a group of acid forming bacteria to simple compounds, such as volatile fatty acids, alcohols, CO_2, NH_3, and H_2.
3. Acetogenesis — the products of the acid-formation are transformed to acetic acid, CO_2, NH_3, and H_2.
4. Methane-formation — acetic acid, CO_2 + H_2 and methanol are transformed to CH_4 and CO_2 by the strictly anaerobic methane forming bacteria.

Depending on the composition of the wastewater, each of the four steps can be rate-limiting.

When the wastewater contains a considerable fraction complex (undissolved particles), the hydrolysis can be rate-limiting; when the pollution consists mainly of acetic acid and other volatile fatty acids, the methane-formation can be the rate-limiting step.

IV. TREATMENT METHODS

The maximal growth rate of the rate-limiting bacteria in the digestion of soluble substrates is in the order of 0.08 to 0.15/day. This implies that the minimal sludge retention time should be 7 to 12 days.

High organic loading rates with medium and low strength wastewaters are only possible when the sludge retention time exceeds the hydraulic retention time. Therefore, the loading rates that can be applied to an anaerobic reactor are primarily determined by the biomass retention of the system. The solution for the biomass retention problem resulted in the development of different anaerobic treatment processes. Schematic diagrams of the various processes are listed in Figure 2.

The essential feature of the ''anaerobic contact-process'' is that the washout of the active anaerobic bacterial mass from the reactor is controlled by a sludge separation and recycle system. The major problem in the practical application of the contact process has always

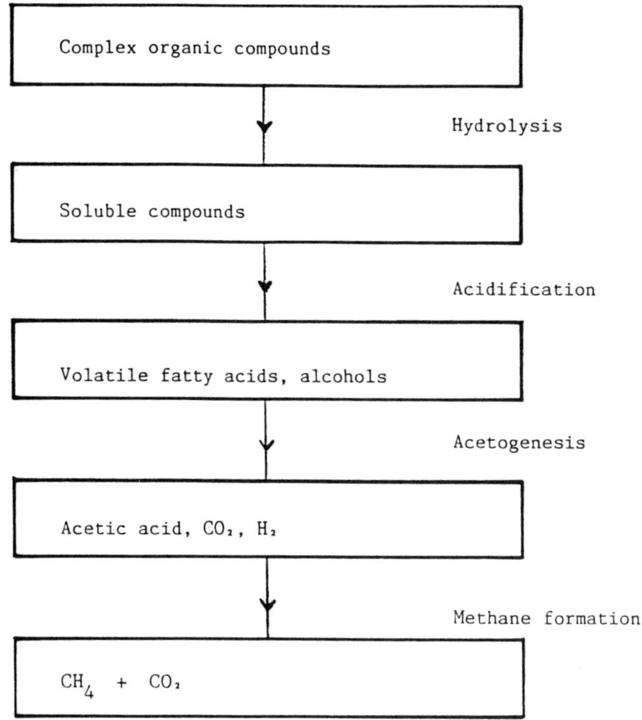

FIGURE 1. Complete degradation of complex organics to methane.

Table 1
HISTORICAL BENEFITS AND LIMITATIONS OF ANAEROBIC TREATMENT OF WASTEWATERS

Benefits	Limitations
Low production of waste biological solids	Anaerobic digestion is a rather sensitive process, e.g., the presence of specific compounds, such as $CHCl_3$, CCl_4 and CN^-
Waste biological sludge is a highly stabilized product that can be easily dewatered	Relatively long periods of time are required to start-up the process when using digested sewage sludge
Low nutrient requirements	Anaerobic digestion is essentially a pretreatment method; an adequate posttreatment is usually required before the effluent can be discharged into receiving waters
No energy requirement for aeration	Anaerobic treatment only seems economically feasible when the influent COD-concentration is high and the temperature is 33 to 37°C
Production of methane, which is a useful end product	Little practical experience has been gained with the application of the process to the direct treatment of wastewater
Very high loading rates can be applied under favorable conditions	
Active anaerobic sludge can be preserved unfed for many months	

been the separation and concentration of the sludge from the purified effluent. Several methods have been used for this purpose: i.e., plain sedimentation, settling combined with chemical flocculation or vacuum degassification, flotation, and centrifugation. The digester contents should be mixed thoroughly by gas recirculation, sludge recirculation, or mechanical agitation.

Another development is the anaerobic filter process. This system simply consists of a

Reactor type	Sludge retention methode	Achievable loading-rates (kg COD.m^{-3}.d^{-1})	
Conventional (completely stirred)	None	Approx. 1	
Anaerobic contact-process	Separate settling tank with sludge-return	Approx. 5	
Anaerobic filter and static fixed film-reactors	Bacterial immobilization on filter material combined with sludge particle retention in filter interstices (upflow or down-flow mode)	10 – 30	
Fluidized Bed	Bacterial immobilization on particles (p.e. sand)	20 – 50	
U.A.S.B.-system	Granulation of bacterial mass and an internal settling compartment	5 – 50	

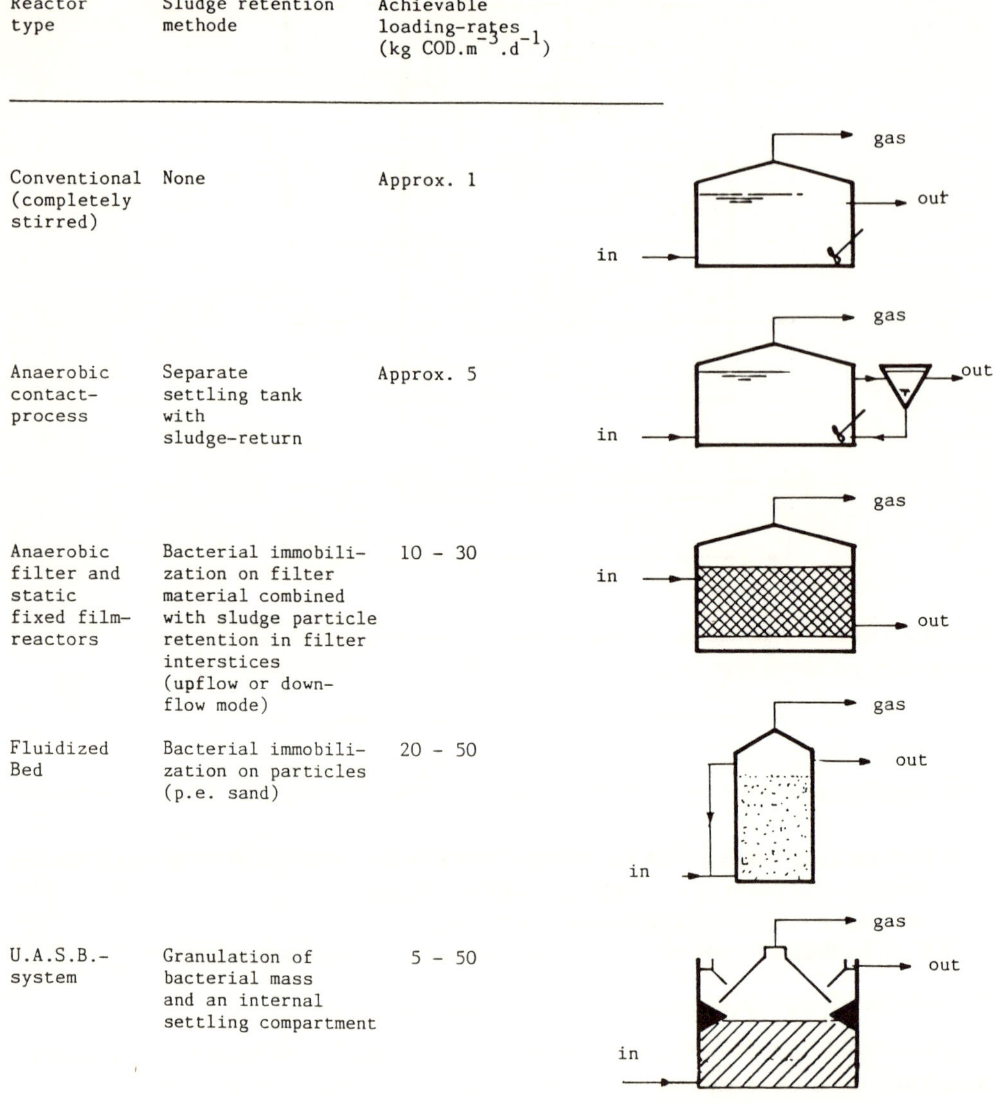

FIGURE 2.   Schematic diagrams and comparison of various anaerobic treatment processes.

vertical filter bed filled with an inert support material such as gravel, rocks, plastic media, or vertical tubes. The sludge is attached on or entrapped in the packing medium.

The anaerobic filter process is suitable to treat various types of chiefly dissolved wastes with a satisfactory treatment efficiency at high hydraulic and organic loading rates. The main disadvantages of the anaerobic filter process are the occurence of scaling (inorganic precipitation) and clogging.

In the fluidized-bed reactor the microorganisms are attached on sand particles (biofilm). To achieve fluidization of the sand bed, a high rate effluentrecirculation is necessary. Influent suspended solids are acceptable to a certain extent, but are not removed.

The U.A.S.B.-reactor is one of the reactor types with a high loading capacity. It differs from other high rate processes by the simplicity of its design. No separate settler with sludge-return pumps are required, as in the anaerobic contact process. There is no loss of reactor volume through filter or carrier material, as is the case with the anaerobic filter and the

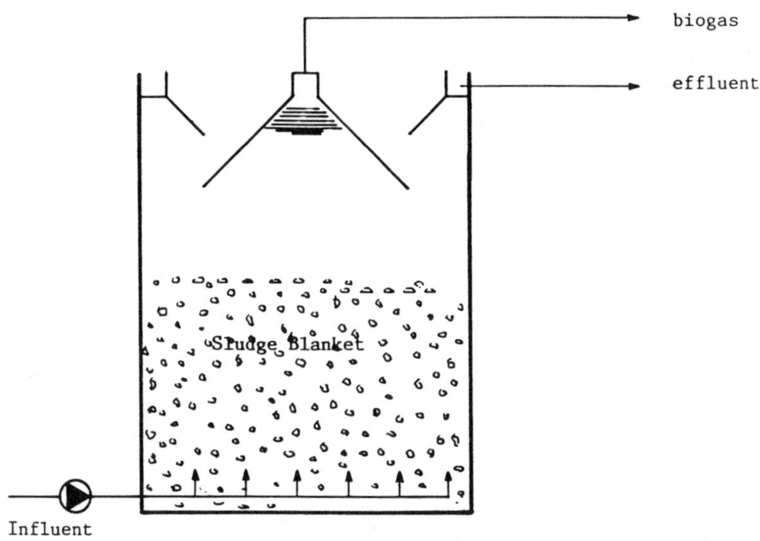

FIGURE 3.   Schematic view of a upflow anaerobic sludge blanket reactor.

fixed-film reactor types. Also, there is no need for a high rate of effluent recirculation and concomitant pumping energy, as is the case with a fluidized-bed reactor.

V. THE U.A.S.B.-REACTOR

The basic ideas underlying the Upflow Anaerobic Sludge Blanket (U.A.S.B.) process are

1.   The anaerobic sludge obtains and maintains superior settling characteristics when the chemical and physical conditions, favorable to sludge granulation and to the maintenance of a well-granulated sludge, are provided.
2.   A sludge blanket may be considered as a more or less separate fluid phase with its own specific characteristics. A well-established sludge blanket forms a rather stable phase, capable of withstanding relatively high mixing forces.
3.   The washout of discrete sludge particles (flocs or granules) released from the sludge blanket can be minimized by creating a quiescent zone within the reactor, enabling the sludge particles to flocculate and to settle.

A scheme of an U.A.S.B.-reactor is presented in Figure 3. The most important features of the U.A.S.B.-concept are the influent distribution system at the bottom of the reactor and the gas/solids/effluent separator, which is installed in the upper part of the reactor. The gas is removed by gas collectors in which a gas/water interface is maintained by a waterlock in the gas outlet. The gas-liquid interface area should be large enough to avoid severe foaming and clogging of the gas outlet pipes. Floating sludge particles must be able to release attached gas bubbles at the interface.

In the reactors space above the gas collectors a gas bubble free-zone permits sludge particles to settle out and to slide back into the digesting compartment along the inclined walls of the gas/solid/effluent separator.[2] Generally no mechanical mixing device is installed in U.A.S.B.-reactors. At high organic or hydraulic loading rates the biogas production and the influent injection guarantee sufficient contact between substrate and biomass.

A properly functioning U.A.S.B.-reactor needs an even distribution of the wastewater at the bottom of the reactor. This is especially true for reactor start-up or during restart after a period of standstill in which the sludge has settled out to a dense mass.

So far the U.A.S.B.-process is the only one of the recently developed processes which is already extensively applied at full-scale. This is especially the case in the Netherlands, but full-scale plants are also under construction or already operating elsewhere in Europe and in the U.S. Since 1979 full-scale plants are in operation for sugar beet wastes, potato processing, potato starch wastes, maize starch waste, and brewery waste. In 1983 the first full-scale plants, treating paper mill effluents were started up and are now functioning very well.

VI. THE U.A.S.B.-REACTOR TREATING PAPER MILL EFFLUENTS

A. Introduction

At Papierfabriek Roermond, a paper mill producing 500 tons of paper daily, an activated sludge plant was built in 1977. Despite the occasionally poor settling characteristics of the sludge, the plant functioned satisfactorily. The increase of the wastewater flow from the start-up time of the plant, made it obvious that the treatment plant had to be extended.

There were two possibilities to be considered: (1) Building a second aerobic plant (parallel); and (2) Building an anaerobic pretreatment plant (U.A.S.B.-reactor). The advantages of an anaerobic pretreatment vs. an aerobic treatment were discussed earlier in this paper.

In September 1981 research was started for anaerobic pretreatment of this paper mill effluent. For this purpose a laboratory-scale U.A.S.B.-reactor (net volume = 30 ℓ) was used. This was the start of a series of experiments that already resulted in the erection of three full-scale plants (operating at design capacity) and the construction of a fourth and fifth plant (start-up: late 1985).

B. Purpose

The purpose of the experiments were to answer the following questions.

1. What level of COD- and BOD-removal can be expected when paper and board mill wastewater is treated anaerobically?
2. What loading rates can be handled?
3. What problems are to be overcome during start-up periods?
4. Will a high-quality granular sludge be developed?

C. Results

The main results of the various experiments are compiled in Table 2. Full-scale results are shown in Table 3. Detailed information about the laboratory experiments and the start-up of the full-scale plant at Roermond are published by Habets and Knelissen.[1] Some aspects of the results are discussed in more details.

1. Start-up

When a reactor is seeded with digested sewage sludge for start-up, the start-up period will last several months (4 to 5 months). During this period, the control and measuring must be quite intensive and the load can be increased slowly.

At the end of the start-up period, the seed sludge has become a granular sludge with high methanogenic activity and good settling characteristics. The reactor can then be loaded at design capacity. When a reactor is seeded with granular sludge from another U.A.S.B.-reactor, the start-up period is much shorter (several days or weeks). From these experiments, it can be concluded that in all reactors a granular sludge was built, independent of the seed sludge. This important result makes it possible to apply high loadings and to reach high removal rates.

Table 2
LABORATORY AND PILOT PLANT RESULTS

Factory/mill	Start-up	Reactor volume (m³)	COD-influent (mg/ℓ)	Temp. (°C)	COD reduction (%)	COD reduction (%)	Biogas production (m³/kg COD-rem.)	Hydraulic retention time (hr)
Paper mill	Sept. 1981	0.03	1600—2000	30	70	—	0.43	2¹/₂
Tissue-factory	Sept. 1982	50	900—2000	30	60—70	—	0.39	2¹/₂
Paper mill De Hoop	Oct. 1983	8	2000	28—33	60—70	85	0.33	4—6
Board mill	Aug. 1984	8	10000	33	80	—	—	—
Paper mill	July 1985	4	2500	32	72	80	0.42	4—6
Pulp mill	July 1985	1.4	20000	35	60	75	0.40	24

Table 3
FULL-SCALE RESULTS

Factory/mill	Start-up	Reactor volume (m³)	COD influent (mg/ℓ)	Temp. (°C)	COD reduction (%)	BOD reduction (%)	Biogas production (m³/kg COD rem.)	Hydraulic retention time (hr)
Board mill BT/NL	Apr. 1983	70	7000	30—37	70	—	0.41	16
Paper mill	Oct. 1984	1000	2500—3500	35	70—85	90	0.42	8
Tissue-factory	Dec. 1984	740	1000	23	60	—	—	6

(2 Weeks After Start-Up)

Factory/mill	Start-up	Reactor volume (m³)	COD influent (mg/ℓ)	Temp. (°C)	COD reduction (%)	BOD reduction (%)	Biogas production (m³/kg COD rem.)	Hydraulic retention time (hr)
Paper mill	Oct. 1985	2200	1200	25	Under construction			
Paper mill	Mar. 1986	1600	2500	32	Under construction			

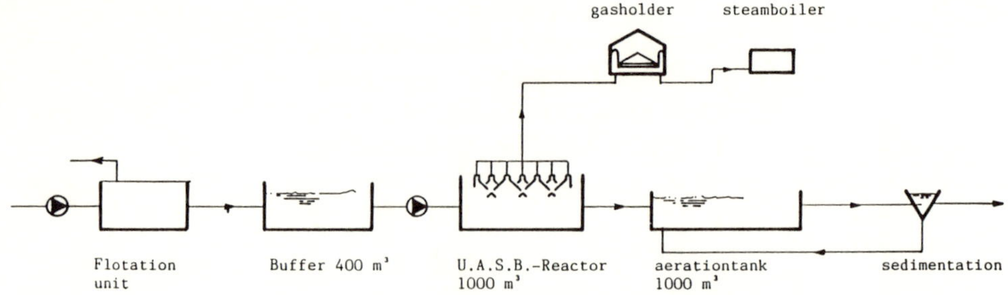

FIGURE 4. Flow sheet wastewater treatment plant.

2. COD-Removal

Average COD-removal percentages are shown in Tables 2 and 3. Removal percentages are fluctuating with influent COD-concentration, while the effluent COD-concentration is quite constant indicating that shock loadings can be handled easy.

3. Biogas

The produced biogas contained 70 to 80% methane, 19 to 29% CO_2, and 1 to 3% hydrogen-sulfide. The carbondioxide concentration is between others influenced by the COD-concentration of the influent and pH of the reactor volume.

4. Volatily Fatty Acids

It was found that 30 to 50% of the wastewater COD was already present as fatty acids, before entering the U.A.S.B.-reactor. The presence of fatty acids (mainly acetic acid) in process water of paper mills is caused by closing-up water systems, and pre-acidification occurs in the sewage system of the mills and in sedimentation units. A good pre-acidification might be a prerequisite for high loaded one phase anaerobic treatment.

VII. THE FULL-SCALE PLANT AT PAPIERFABRIEK ROERMOND

A. Description of the Plant

Figure 4 shows a flow sheet of the complete anaerobic/aerobic plant. Surplus of clarified process water is discharged at a temperature of 35°C into the biological treatment plant. A part (400 m³) of an existing aeration tank is used for wastewater buffering before transportation to the U.A.S.B.-reactor takes place. The anaerobic effluent flows down to the aeration basin, where purification is completed.

The reduced load on the activated sludge plant brought several advantages. After upgrading, sludge settlement improved, final effluent quality improved, and an important amount of electricity is saved because less air diffusion is required. One of the existing aerobic basins is out of use, the other is partly used as a buffer (400 m³) and as a posttreatment (1000 m³).

B. Cost Analyses

Upgrading the effluent treatment plant at Papierfabriek Roermond was necessary, because the existing plant was already overloaded. Before the decision was made, cost calculations were made for the present situation and two alternatives:

Table 4
ANNUAL COSTS IN $1000 FOR
THREE ALTERNATIVES FOR
WASTEWATER TREATMENT AT
PAPIERFABRIEK ROERMOND

	A	B	C
Interest (10%)	120	180	160
Depreciation (15 years)	80	120	107
Electricity	70	135	33
Chemicals	40	45	45
Maintenance	20	30	25
Labor	20	20	10
Sewer	10	10	10
Total	360	540	405
Gas proceeds	—	—	125
Net cost	360	540	280

Note: A = no upgrading (40,000 in/hr equivalent); B = Extension of aeration capacity (80,000 inh. equivalents); and C = Anaerobic pretreatment + steam production (80,000 inh. equivalents).

1. Present situation: capacity is 40,000 inhabitant/hr equivalent to (7.5 tons of COD/day); investment in 1977 was $1,200,000.
2. Upgrading the aeration capacity to 80,000 inhabitant/hr equivalent to 15 tons of COD/day; investment $600,000.
3. Upgrading to 80,000 inhabitant/hr equivalent to anaerobic pretreatment and gas utilization in a steamboiler; investment $330,000 and $70,000 respectively.

Table 4 shows the different costs. Alternative B was expected to be $180,000 per year more expensive than before extension, mainly due to higher capital and electricity cost. Alternative C showed lower investment costs than B and even lower electricity costs than A. Thus, it would be only $45,000 more expensive than the original situation A. After subtracting gas proceeds it appears that alternative C is even more favorable than without upgrading.

VIII. CONCLUSIONS

It has appeared from experiments and full-scale plants that anaerobic treatment of paper and board mill effluent according to the U.A.S.B.-concept is very promising. The following conclusions can be drawn.

- Anaerobic pretreatment is a very attractive method for extending an aerobic plant.
- It can be expected that application of anaerobic treatment will strongly extend during the next years.
- Start-up of a full-scale plant can be completed in about 4 months or in case adapted granular sludge is available in several weeks.

REFERENCES

1. **Habets, L. H. A. and Knelissen, J. H.,** Application of the U.A.S.B.-Reactor for anaerobic treatment of paper and board mill effluent, *Water Sci. Technol.,* 1985.
2. **Lettinga, G., van Velsen, A. F. M., Hobma, S. W., Zeeuw, W. de, and Klapwijk, A.,** Use of the upflow sludge blanket (U.S.B.) reactor concept for biological waste water treatment, especially for anaerobic treatment, *Biotechnol. Bioeng.,* 22, 699, 1980.
3. **Zeeuw, W. de,** Acclimatization of Anaerobic Sludge for U.A.S.B.-Reactor Start-Up, Ph.D. thesis, Agric. Univ. of Wageningen, Netherlands, 1984.
4. **Swinkels, K., Vereijken, T., and Hack, P.,** Anaerobic Treatment of Wastewater from a Combined Brewery, Malting and Soft-Drink-Plant, 20th Int. Congress European Brewery Convention, Helsinki, June 1985.

Chapter 6

BIOENERGY RECOVERY AND CONSERVATION IN ISRAELI AGRICULTURE — ANAEROBIC DIGESTION OF COTTON STALKS

Uri Marchaim and Carlos Dozoretz

TABLE OF CONTENTS

I. Introduction ... 156

II. Conservation of Energy in Greenhouses ... 156

III. Conservation of Energy in Animal Husbandry ... 157

IV. Energy Recycling in Animal Production by Improving Feed Digestibility ... 157

V. Agriculture Waste Utilization in Israel ... 158

VI. Composition of Plant Materials ... 160

VII. Chemical Composition of Cotton Stalks ... 161

VIII. Anaerobic Digestion ... 162

IX. Anaerobic Digestion of Cotton Stalks ... 163

X. Thermophilic Anaerobic Digestion — Preliminary ... 164

XI. Thermophilic Anaerobic Digestion — Advanced Experiments ... 165

XII. Mesophilic Anaerobic Digestion — Simple Batch Digestion ... 167

XIII. Discussion ... 169

XIV. Conclusions ... 172

References ... 173

I. INTRODUCTION

Modern agriculture is highly dependent on other sectors of the economy for most of its inputs. Agricultural development in the last years showed that the rate of dependency of agriculture on other sectors is increasing. This is noted especially in the growing demand for fossil fuel energy. Modern agriculture cannot greatly reduce its demand for energy without a major reduction in output.

Israeli agriculture is probably one of the most energy-intensive industries in Israel and in World agriculture. Its increasing productivity and development had involved increased energy use. In a recent study by Dvoskin et al.,[1] the impact of rising energy prices on various agricultural sectors was evaluated. They found that if energy prices will continue rising the agricultural economic situation deteriorates faster than the industrial sector.

Despite its energy intensity Israel has managed to utilize its natural resources for agriculture. For instance, the NEFAH[2] project — the utilization of agricultural wastes and biomass for production of energy (biogas), animal food supplement, substrates for crops and plants, and industrial products (fiber boards) by the anaerobic digestion process — is unique in its comprehensive approach and is beginning to be widely used in Israeli kibbutzim.

In Israel over 10% of the total national energy consumption is used for agricultural production, and only about 2.5% in the form of direct energy input.[3] The profitability of many crops depends very much on the prices of fuel and its availability. The high energy intensity of Israeli agriculture is mainly due to its water needs and the constant growing of the agricultural production for export, mainly to the European market. Therefore, reduction in energy consumption, and use of alternative energy sources are in very intensive development in Israel.

Two ways to ensure that agriculture has the energy that it needs are

1.   Substitution of plentiful energy agricultural users with other energy sources.
2.   Conservation and more efficient energy management.

A brief description of the main efforts in Israel will be described in this paper with a deeper description on our specific work with cotton stalks.

II. CONSERVATION OF ENERGY IN GREENHOUSES

Air shipment, greenhouses heating, artifical lighting, and fertilizers makes this exported crop very energy dependent. More than half of the production cost of cut flowers is for energy. Therefore, reduction in energy consumption and use of alternative cheaper energy sources are of main goals in the Israeli agriculture.

In a symposium held March of 1984, at the Volcani Center at Beit-Dagan Israel, the energetic aspects of growing plants undercovers (greenhouse and plastic covers) were discussed and summarized.[4] This followed a survey of greenhouse heating systems in Israel done by Heshev.[5]

It is clear that the plant does not need energy at night. We heat the greenhouse since the loss to the environment is high because of the greenhouse structure which is a compromise with the needs of the flowers during the day and summer. During the summer days we even need to lower the temperature in the greenhouse. Therefore, most new technologies for new structures of greenhouses or new ways of heating are based on insulating the greenhouse, collecting the surplus heat during the day, and releasing it during the night or even during some of the winter days.

In Israel, several new technologies were examined,[4] including closed greenhouses in which air changes are limited, uses of hygroscopic salts, developing heat exchanger light/air on the roof, and the hydrosolaric greenhouse system. In all those directions a very intensive research and development is done and results are implemented in many places even before final results are published.

III. CONSERVATION OF ENERGY IN ANIMAL HUSBANDRY

Until a few years ago not much was done by Israeli agricultural section to reduce the negative impact of energy soaring prices on the farming community. However, in the last few years, more and more farmers have been taking an interest in solving their energy problem. The government is helping in financing some of those efforts.

Because of the special nature of the energy consumption pattern of Israeli animal husbandry, not much direct energy can be saved. Most energy saving will be achieved through better use of inputs. The following are some examples of energy saving:

1. Instruments for better control of the temperature inside poultry houses, and using the heat of the chickens themselves in a more efficient way.
2. Reducing energy use by better design of equipment.
3. Reducing energy demands for hot water by using waste heat from compressures of the milking system.
4. Reducing energy use of water pumping which is the main sector consuming energy in Israeli agriculture.
5. Reducing energy consumption in fish ponds by reducing airation.
6. Reducing indirect energy of fertilizers, pesticides, and irrigation system operations.

IV. ENERGY RECYCLING IN ANIMAL PRODUCTION BY IMPROVING FEED DIGESTIBILITY

Food for cattle and poultry has to be looked at as metabolic energy, when dealing with energy conservation and better utilization. It is important to pay attention to processes preparing the feed and uses of the animal wastes as materials for refeeding. If animal feed is considered as metabolic energy, improving digestibility of organic materials will have an important role in energy conservation and utilization. This will also decrease agricultural inputs by using less water, fertilizers, and direct energy.

In the last few years, more and more farmers and researchers have been taking an interest to find alternative waste materials for feeding animals. It is a common procedure in most Israeli animal husbandry to use a computerized program that has the ability to price the feed based on the nutritional data provided according to a lesser cost gain and profit projection program. This program can compare the waste provided to the existing value of other conventional energy and protein feedstuffs and determine the amount that can be used. Different waste materials like citrus peels, cotton seeds, grain dust, whey etc., are widely used in Israel. Other organic wastes are specially treated.

Cellulose is, as might be expected, one of the principal sources of energy in ruminant nutrition due to its fermentation in the rumen by cellulolytic microflora. Notwithstanding, the diet of a highly productive ruminant, under intensive conditions, is characterized by a high content of readily fermentable carbohydrates and is low in cellulose. This dietary regime should be maintained in order to keep the high level of production of the ruminant industry. However, as the prices of the starchy feeds are high and increasing continuously, the time has come to consider the conversion of cellulosic materials into readily fermentable feeds. In considering prospective uses for cellulose, this direction appears to be particularly attractive, from both the economic and technological points of view, since the modest objective (according to this concept) is to produce a crude fermentable mixture.

Ben-Ghedalia et al.[6] examined the effect of combined chemical and enzyme treatments on the saccharification and in vivo digestion rate of wheat straw. Wheat straw was pretreated with sodium hydroxide, ozone, and sulfur dioxide, and subsequently treated with four sources of cellulases. The effect of the combined chemical plus enzymatic treatments on the extent of saccharification and on the digestion rate by rumen microorganisms was studied. The in vitro organic matter digestibility was affected significantly only by the chemical pretreatments, whereas the effect of the cellulases was expressed mainly in increasing the fermentability of the hydrolyzed straw. The in vitro digestion pattern of the saccharified straw was found to be typical of a highly fermentable feed comparable to a starchy mixture such as that used in concentrate ruminant diets.

Ben-Ghedalia et al.[6] also examined the effect of ozone and ammonium hydroxide treatments on composition and in vitro digestibility of cotton straw. In vitro organic matter digestibility was significantly increased by more than 100% by the ozone treatments, as a result of the partial conversion of cell walls into cell content and the increased degradation of the cell wall. Ozone treatments increased the initial rate of the in vitro organic matter digestibility.

Those treatments are still under intensive research and are not widely used. Work done with the feeding of raw cow manure[7] is also examined but not used. However, poultry manure (after silage for cattle or as is for fish) is a common feed in Israel and was examined for years by Tagari[8] and colleagues of the Faculty of Agriculture in Rehovot.

V. AGRICULTURAL WASTE UTILIZATION IN ISRAEL

Plants transform solar energy into chemical energy or biomass, which can then be harvested and converted into usable energy products. The biomass resources of residues and terrestrial and aquatic plants are renewable; they can potentially supply significant amounts of the future energy requirements. In the historical pattern of agriculture, crop residues went back to the land, by ploughing it back to the soil.

Since about 1970, many countries in the world has undertaken an extensive research program to develop processes for converting agricultural, foresty, and municipal wastes into fuels and chemicals. This program has involved studies on both old and new processes for biomass conversion. Many of the older processes of biomass conversion, such as acid hydrolysis and fermentation to ethanol and to biogas, are being reinvestigated using newer technologies.

These resources can be converted by a number of technologies into liquid transportable fuels, gaseous energy products, and other forms of energy. The present availability of biomass residues (animal manures, field crop, forest, and mill residues) is estimated to be 100 million tons dry matter, with a theoretical maximum of 427 million tons dry matter.[9] Biomass residues are currently being used in animal feeds, fiberboard products, fertilizer, soil tilth, and as soil conditioners.

The cost of biomass residues varies greatly according to type, application, region and location with respect to potential users. Field crop residues vary in the U.S. from $4/ton for wheat and other small grain straws to $15/ton for corn stover. Manure residues range from $0.50/ton for cattle manure to $17/ton for poultry manure. The cost associated with collecting, reducing, and transporting forest residues is estimated to range from 25 to $60/ton, depending on location.

Many processing options can be applied to the conversion of biomass to energy or chemicals. These processes range in their stage of development from laboratory scale to commercially proven processes. Commercial combustion facilities have been used for many

Table 1
OVERALL QUANTITIES OF
AGRICULTURAL WASTES IN
ISRAEL[10]

Type of waste	Inclusive quantity of dry organic matter (tons)
Manures	745,000
Crop waste	870,000
Agro-industrial waste	500,000

years, and anaerobic digestion facilities for feedlot manures are operating now. Fermentation to produce ethanol and petrochemical substitutes, gasification to produce ammonia, SNG, hydrogen and methyl fuel, and liquefaction to produce fuel oil are technically feasible but not yet economically competitive. Photochemical processes are being pursued on a long-term low emphasis basis.

Some biomass materials may require pretreatment, which usually consists of drying or further volume reduction. After suitable preparation, biomass can theoretically be processed using any of a number of proven technologies. However, the chemical composition of any biomass feedstock will dictate the probable conversion process and path for utilization. For example, direct combustion or gasification is the most likely alternative for low moisture content biomass, while anaerobic digestion and fermentation are more adaptable to high moisture biomass materials.

There are three major sources of residual agricultural biomass in Israel: manure, crop, and agro-industrial wastes. The relative amounts of each type of biomass were published earlier[10] and are summerized in Table 1. In a survey published by Marchaim and Criden[10] and a survey published by Rymon and Wienstein[11] it was found that over 870,000 tons of dry matter of organic crop wastes are generated every year in Israel. The largest part of this waste is generated at the cotton fields — over 260,000 tons of dry weight per year. The other main waste is the wheat straw (250,000 tons). In another survey conducted by TAHAL Engineering Co.[12] the amounts differ slightly and an economical evaluation of some technologies to produce energy from biomass are presented.

Most, if not all, of the cotton waste is not utilized yet. The main reasons for that are

1. Problems in storage of this waste, which is generated in high quantities in a very short period
2. The high moisture (50 to 65%) which require drying or special storage, in order to stop biochemical processes
3. Lignification during storage which lower its quality for feeding
4. A very low quality as animal food without special pretreatment, which is expensive
5. High levels of chemicals used during the growing season

A preliminary study, done by us a few years ago,[13] examined the possibility of using cotton stalks as raw material for growing yeasts with a high methionin content, as animal feed. The study examined the distribution of cotton stalks in Israel and the availability of this waste. It was found that in some parts of Israel the process can be economical feasible.

It has been estimated[14] that the biomass of waste could replace 5 to 10% of the petroleum need in the U.S., and we estimated[10] that in Israel 8 to 10% of the current used energy could probably be replaced.

Much activity in biotechnology has been directed toward the use of renewable plant resources for production of fuels and energy. Another important use could be for production

Table 2
APPROXIMATE COMPOSITION OF SOME
CELLULOSIC MATERIALS (%)[17]

Material	Cellulose	Hemicellulose (% dry weight)	Lignin
Coniferous wood	40—50	20—30	25—35
Deciduous wood	40—50	30—40	15—20
Corn cobs	45	35	15
Corn stalks	35	23	35
Wheat straw	30	50	15

of industrial chemicals. A route of great potential for basic chemical production of sugars from plant biomass is via hydrolysis of the carbohydrates. Simple carbohydrate and structural polymers represent over 80% of forage biomass.[15]

The depolimerization products of fiber components must be separated one from another in order to convert them selectively to a mixture of simple organic compounds. Hemicellulose sugars are released following mild acid hydrolysis. The more resistant cellulose can then be hydrolyzed to glucose by more severe acid hydrolysis conditions. Structural polysaccharides from mature crops needs high temperature steam treatment which are very costly although not very effective.

The discussion to follow will not dwell on the hydrolysis processes, which have recently been reviewed by several authors in the *CRC Series in Bioenergy Systems,*[16] but will concentrate on the anaerobic digestion of crops to produce biogas and other useful products.

VI. COMPOSITION OF PLANT MATERIALS

Most plants materials contains three major components: cellulose, hemicellulose, and lignin. Cellulose is generally the major component, making up 25 to 61% of all plant materials.[17] The other components, hemicellulose and lignin, differ in their structure, depending on the plant material. The relative abundance of these components in various biomass materials are shown in Table 2.

The structure, origins, and potential applications of lignin are considered in the light of the chance that lignocellulose materials may one day emerge as significant sources of biomass based energy production. If the cellulosic portion of such materials are used to produce energy, then the utilization of the remaining lignin assumes an important role and has related economic implications.

Cellulose is the single most abundant renewable resource on earth. It has been estimated that as much as 22×10^9 tons of cellulose are produced annually worldwide through the photosynthetic fixation of CO_2 and that as much as 4×10^9 ton/year could be available for processing. In nature, cellulose, a polymer of β-D 1,4-linked anhydrous glucose units, comprise 40 to 60% of the cell wall material of trees and plants. The individual cellulose molecules are linked together to form elementary fibrils, 30 to 40 Å. These fibrils are aggregated, via intermolecular hydrogen bonding, into larger subunit structures referred to as microfibrils. Of essentially infinite length and approximately 250Å width, these microfibrils contain alternating sequences of highly ordered (or crystalline) and randomly oriented (or amorphous) phases and are imbedded in a matrix of hemicellulose.

This latter carbohydrate component, hemicellulose, comprising 20 to 50% of the plant dry weight, is a branched polymer of pentose sugars. The composition of a specific hemicellulose of hardwoods differs from that of softwoods, and those will probably be completely different from the cotton stalks hemicellulose.

In mature plants, the cellulosic and hemicellulosic fractions are encrusted in an amorphous lay of lignin. This latter component, which accounts for the remaining 10 to 20% of the plant material, is a complex three-dimensional polymer formed by carbon-carbon or ether bonds between phenylpropane units. It is the intricate nature of the association of cellulose with hemicellulose and lignin and of the intermolecular hydrogen bonding of cellulose itself which simultaneously supplies structural integraty to the plant and hinders the access of degrading agents to the cellulosic portion of the biomass. Native biomass can be roughly thought of as similar to reinforced concrete with crystalline cellulose fibers (analogous to reinforcing roods) imbedded in a matrix of lignin (analogous to the concrete). The lignin also serves to seal the cellulose, forming a barrier which protect the fiber from attack by hydrolytic agents.

Effective utilization of this renewable resource depends in part on its efficient hydrolysis to yield glucose. Cellulose is resistant to hydrolysis either by acid or enzymes. Furthermore, cellulose naturally occurs in association with hemicellulose and lignin. Recently, a number of pretreatments have been proposed to disassociate the lignocellulose complex (the Iotech process of steam explosion, enzymes, or acids and a pretreatment by ozone). Commercialization of such systems to produce glucose syrups is envisaged.

Acid hydrolysis of cellulose, though rapid, has a number of disadvantages which include: high capital costs, low overall conversion efficiency (65 to 72%), degradation of product (glucose) to yield toxic impurities, and equipment corrosion. On the other hand, enzymatic hydrolysis of cellulose yields pure glucose syrups devoid of degradation products and only requires low temperature and pressures. It is far more efficient in terms of percentage conversion (95%). However, the enzymatic process is slow. The lignin is almost not degraded in the regular enzymatic process, but interfere with the penetration of the enzymes.

The high amount of the cotton stalks waste, containing high percentage of lignocellulose complex, is a target for many chemical as well as biochemical treatments in addition to the direct use for burning (a project now in final stage of eraction at Sha'ar Hanegev, Israel).

The chemical procedures are used in many cases as pretreatment for the biochemical processes. It was examined in the U.S. by several researchers and is also examined now in Israel,[13] for a process of growing SCP (Single Cell Proteins). Until recently, those processes showed no economical profit and in our work we are looking for a process that will avoid the pretreatment. One of the possibilities is the thermophilic anaerobic digestion that already showed some advantage in degrading fibers.[19]

VII. CHEMICAL COMPOSITION OF COTTON STALKS

We define cotton waste material as the part of the plant that remains in the field after the picking of the cotton and the cutting down of the stalks. A stalk of approximately 10 cm is left and the roots are not uprooted. The stalk waste was found to be composed of 50% main stalk, 30% small stems, 6 to 10% leaves, and 10% cotton fibers.[20] In several laboratories it was found that the cotton stalk differs from wheat straw in some chemical features.

1. The lignin percentage ranges betweed 12 to 22% in cotton stalks compared to 9 to 12% in wheat straw. The lignocellulose in cotton stalks is also more complex and harder to break, chemically or enzimatically, than other types of straw.
2. The hemicellulose in cotton was about 13% while it was approximately 25% in wheat straw. This means a lower portion of degraded organic matter after hydrolysis in the cotton stalks. The cellulose content of cotton stalks was slightly higher (30 to 45%) than in wheat straw.[18]
3. The cotton stalks contain approximately 4 to 7% pectin compounds which are neglectable in wheat straw.

Table 3
CHEMICAL COMPOSITION OF STORED COTTON STALKS,
HARVESTED ON SEPTEMBER, DRIED IN THE FIELD AND
KEPT IN CLOSED CONTAINER

Date	Solids (%)	Ash (%)	C.O.D. (g/kg)	Total nitrogen (g/kg)	Ammonia (g/kg)	pH	Volatile acids (g/kg)
February	87.4	8.0	166.95	2.4	0.1	7.0	9.696
March	90.6	6.0	307.77	2.0	0.1	7.0	7.488
April	88.7	—	505.92	1.7	0.1	7.0	—

4. The cotton stalk contains 20 to 30% intracellular soluble materials while the wheat straw contains only around 13%. This difference shows the potential of cotton stalks as a source for extracting the plant metabolites, many with biological activities.
5. The cotton stalks contains soluble ash, while in wheat straw the ash was mainly insoluble composed of silica.

It was found[18] that the composition of the cotton stalks depends on the quality of the soil it was grown on. Specifically in the Hula Valley fields with the high water levels and high organic matter, it was found that the water content is much higher in the cotton stalks than in other parts of Israel.

A more detailed information on the components of the cotton stalks in several parts of Israel can be found in the work published by Ben-Ghedalia et al.[6,20] of the Volcani Institute.

It was found in the work we did earlier at the Migal laboratory that the chemical composition of cotton stalks harvested on September, dried in the fields and then kept for a few months in a closed container of 1 m^3 is not constant. Since there is an effect of heterogenity and the material collected from the field consist of the stalks, leaves, and some cotton left, the exact composition depends on the specific material in a specific location. It also depends on the period of storage, and the longer the storage the higher the lignin. Results are given in Table 3.

VIII. ANAEROBIC DIGESTION

Anaerobic digestion of organic matter has been implemented and recommended for many years as a practical process for disposal or energy recovery, mostly for sewage treatment. Intensive utilization of the technology as a source of energy has not yet occurred. Now, the increasing value of fuels and the impending shortage of the more convenient fuels, notably natural gas, creates a climate in which anaerobic digestion becomes of strong interest. Evidence of this interest was work carried out in the last few years to provide the technical basis for generating fuel gas by the anaerobic digestion of municipal solid waste.[22] Another resource for fuel gas generation is the residue from intensive animal raising operations such as beef cattle feedlots and dairying.[10]

The publication of Buswell[23] at the University of Illinois deserves recognition as a landmark in anaerobic digestion technology. The agricultural residues of chief concern to Buswell were plant materials; he studied the digestion of pure cellulose as well as many other representative pure compounds in order to understand what ultimate performance might be attained in digesting residues and to determine how well corn stalks could be digested. It was determined that all the water soluble constituents are removed in digestion and that the chief water insoluble material attacked was cellulose. The experiments showed that some lignin was attacked, but in other work he demonstrated that isolated lignin does not digest and, in fact, suppresses the digestion of cellulosic materials through the recovery of fibers for paper making.

Anaerobic digestion is a process wherein complex organic compounds are broken down to produce biogas (consisting of methane and carbon dioxide). It takes place in the absence of air and in the presence of suitable population of microorganisms. Temperature is a basic parameter, and the speed of the reaction can be increased by raising the temperature. Two optimum temperature ranges — mesophilic (30 to 40°C) and thermophilic (50 to 60°C) — are usually used, and differ in their bacterial population. Important environmental factors in digester management include temperature, pH, nutrient supply, absence of oxygen, and toxic materials. During the anaerobic digestion process the biodegradable fraction of the organic matter fed to the digester is attacked by the microorganisms and the degraded matter is converted to biogas which can be used as a fuel. The undegradable matter (the percent of which depends on the digestion process and condition as well as on the fraction of cellulose and lignin) consists mainly of the fibers.

The functioning and management of anaerobic digesters can be approached from standpoints which differ in level of detail. The most general outlook is an engineering approach, establishing good operating principles. A more detailed approach deals with the extracellular chemistry of the digester. This focusses on the more readily measurable chemical characteristics of the system. At the most detailed level, the fundamental microbiology of anaerobic digestion is considered. This approach has the promise of yielding the broadest understanding, but the scope of information available is not yet complete enough to provide an easy basis for digester manipulation. A number of reviews of anaerobic digestion technology are available based on approaches of this kind. As examples, there may be cited the practical summaries by Buswell.[24] McCarty[25] presents the chemical approach. The review by Hobson et al.[26] stresses the metabolic mechanisms.

The anaerobic conversion of a substrate to methane is commonly referred to as occurring in two stages, acid formation and methanogenesis. Two groups of microorganisms mediate these two stages of digestion.

IX. ANAEROBIC DIGESTION OF COTTON STALKS

Many different organic wastes can be used as biomass sources for feeding anaerobic digesters. The integrative approach to the subject of biomass production and utilization raises the demand to reconsider the introduction of biomass from sources which were previously neglected. With this approach, the collection of different wastes (especially plant residues) which were formerly abandoned and even cultivation of biomass as energy crops to increase the amount of substrate of the process, may become feasible.[27]

Choosing the optimal biomass source for an integrative project is a complex task requiring a comprehensive survey of agroeconomic factors, e.g., plant growth rates, land, water, and fertilizer demands and availability, etc. Anaerobic biodegradability factors should be considered, as well as the quality of digested products for further potential utilizations at the specific locality.

A potential benefit of using biomass from different sources is the possibility of controlling the chemical composition of the digester feed. Of special importance to the digestion efficiency is the balanced ratio of N, P, and C, but some other elements also play an important role in maintaining an effective digestion process. Organic matter of plant origin is often characterized by the C:N ratio, which is too high for anaerobic digestion, whereas the ratio of these same elements in municipal sewage sludge and in livestock wastes (especially swine and poultry manures) typically is low.

Cotton stalks alone were not examined for anaerobic digestion since it is known from our experiments in the past and from work done by others that this waste needs some other organic material to be mixed with in order to initiate any anaerobic digestion activity.

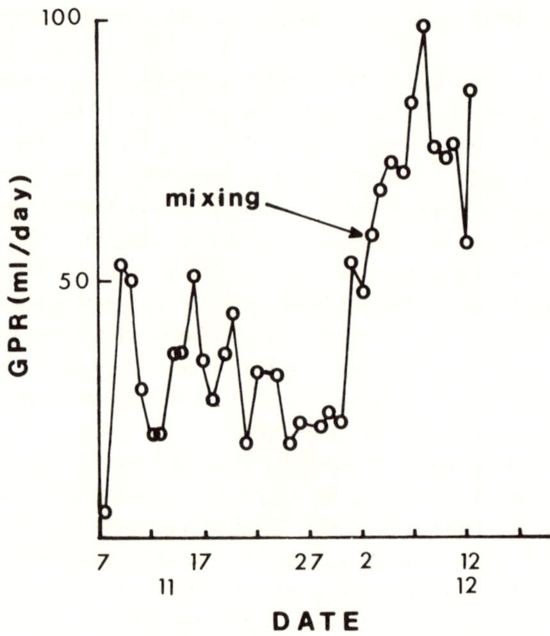

FIGURE 1.   Biogas production rate in thermophilic anaerobic digester (of 2 ℓ working volume) of cotton stalks without stirring.

X. THERMOPHILIC ANAEROBIC DIGESTION — PRELIMINARY

The advantages of using the thermophilic anaerobic digestion for the breakdown of crop waste was presented by Marchaim et al.[19] recently. It suggested that when agricultural wastes, and especially crop wastes, are used as raw materials, the thermophilic digestion has an advantage. The main reason for that assumption is the different quality of the digested material left after the anaerobic process which became the main product of the process, and is called "Cabutz".

Cabutz is the solid phase of the slurry left after the thermophilic anaerobic digestion. It consists mainly on the lignocellulose complex of the organic material charged to the anaerobic digestion system, which was degraded to a limited extent. It was shown that the Cabutz is a growth substrate for house plants and can replace peat moss in the greenhouses[28] and mushroom industry.[29] This material is now used in many places in Israel and has proved to produce high yields of plants; it probably contains some growth factors. In the NEFAH process, developed in Israel, the thermophilic anaerobic digestion of cow manure not only broke down the lignocellulose complex to achieve the porosity needed for air penetration and water-holding capacity, but also eliminated most pathogenic bacteria present in the manure[30] and most moulds.[29] The Cabutz therefore has some advantages over peat moss when used in the integrative farm, since it is less likely to become contaminated.

The anaerobic digestion was performed in an erlenmyer of 3 ℓ, containing up to 2 ℓ of water in which cotton stalks were imersed and digested anaerobically in a water bath heated by thermostate of 53°C. The material in the vessel was stirred with a magnetic stirrer (which sometimes, especially with high solids content, did not work properly). The gas was measured by water displacement. Cotton stalks of the last harvesting period were collected in the field after trimming and kept in a closed container (Table 3). The stalks were grounded and mixed with water to the proper solid concentration. The gas production rates (GPR) of both unstirred and stirred digesters are shown in Figures 1 and 2 respectively.

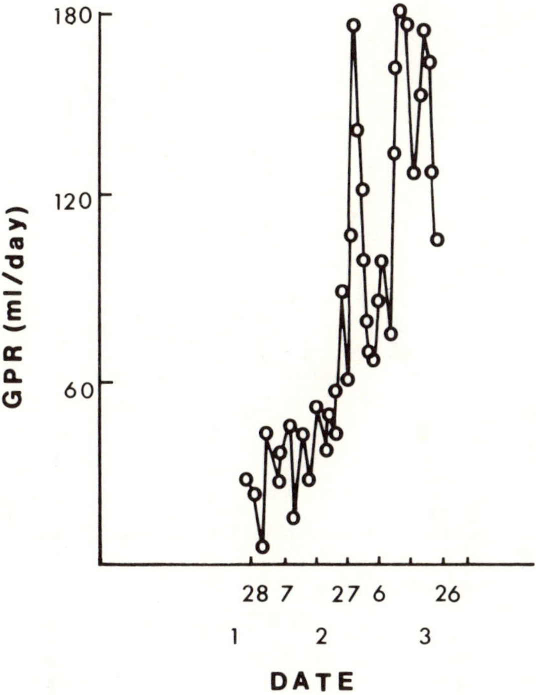

FIGURE 2.   Biogas production rate in thermophilic anaerobic digester (of 2 ℓ working volume) of cotton stalks, with stirring.

Another digester used for the anaerobic digestion of cotton stalks was a perspex vessel of 6.5 ℓ (4 ℓ working volume) with mechanical stirring at 60 rpm in a water bath of 53°C. There was a 2.5 cm outlet at the bottom of the vessel for slurry withdrawing and two 1 cm pipe openings for feeding and biogas. The main problem in this system was not in feeding the materials through the 1 cm pipe but the withdrawing of the representative sample from the 2.5 cm outlet. Although the stirrer was in operation, only the liquid fraction was withdrawn, and accumulation of the fibrous fraction was noted in the digester. Therefore results of this experiment are only an indication. The gas production rates are given in Figure 3.

The pH of digestion was around 7.6 and volatile acids concentration was lower than 2 g/ kg. No results of the chemical composition of the withdrawn slurry during the experiment is included since the material withdrawn was not representative.

During the accumulation of solids in the digester, the growing levels of the volatile acids probably affected the pH and souring of the system began. At the end of the experiment in the 4 ℓ working volume digester (6.5 ℓ total volume) samples from the upper, middle, and lower parts were collected and analyzed. Results are given in Table 4.

The low pH explains the low gas production, and the differences in solids concentration between the different layers — the problem in mixing and in gas production.

XI. THERMOPHILIC ANAEROBIC DIGESTION — ADVANCED EXPERIMENTS

In a work done by Kimchie of the Haifa-Technion group,[31] as part of the NEFAH project, the contribution to gas production of cotton stalks in mixture with cow manure in a 12% solids concentration in the digester was examined. The percentage of the cotton stalks dry

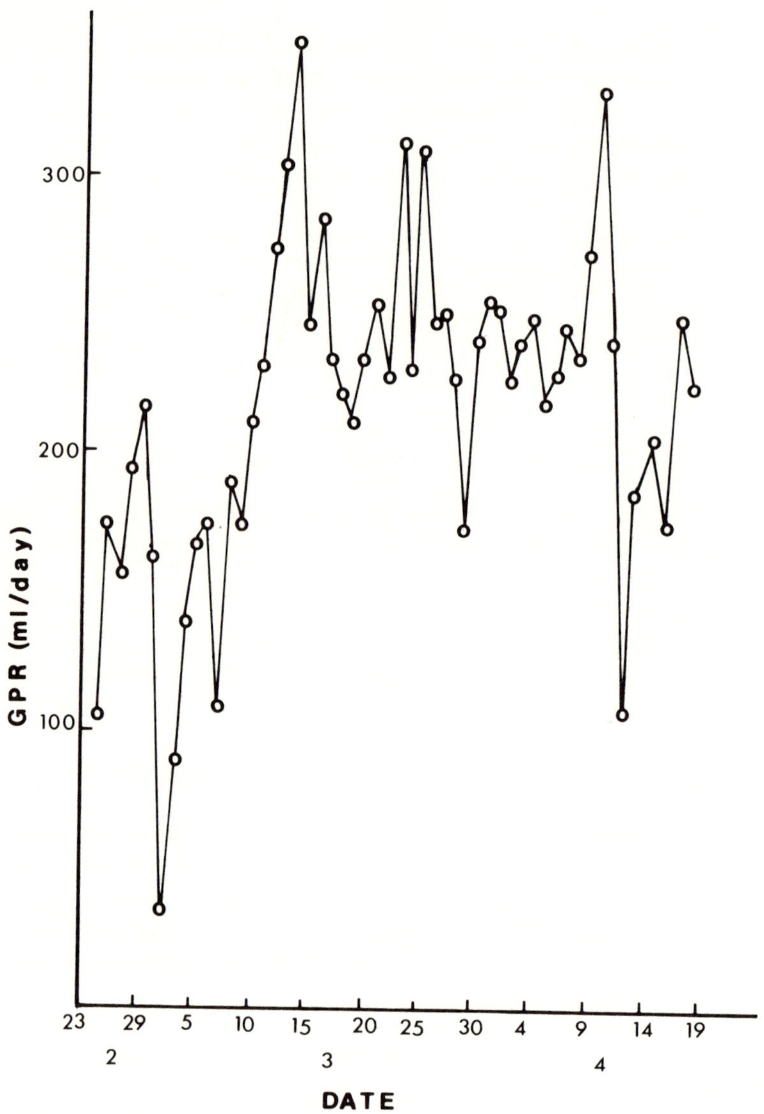

FIGURE 3.   Biogas production rate in thermophilic anaerobic digester (of 4 ℓ working volume) of cotton stalks, with mechanical stirring. The digester was charged continuously once a day with powdered cotton stalks immersed in water.

Table 4
CHEMICAL COMPOSITION OF DIFFERENT LAYERS OF THERMOPHILIC DIGESTED SLURRY AT THE END OF EXPERIMENT 2

Layer	Solids (%)	Ash (% dry matter)	COD (g/kg)	Total nitrogen (g/kg)	Ammonia (g/kg)	Phosphate (g/kg)	pH
Upper	17.48	13.0	111.30	5.5	4.6	1.0	6.3
Middle	13.44	19.0	733.58	5.7	5.1	1.0	6.5
Lower	6.95	9.0	383.65	9.6	4.0	2.6	7.0

Table 5
EFFICIENCY OF ANAEROBIC DIGESTION
OF COTTON STALKS IN MIXTURE WITH
COW MANURE IN THERMOPHILIC
DIGESTION WITH RETENTION TIME OF 10
DAYS AND 12% SOLIDS IN THE DIGESTER[31]

TS cotton of mixture	Gas production rate	Digestion efficiency in feed	
		Manure solids	Cotton solids
(%)	(ℓ/ℓ/day)	(ℓ/g solids)	
0	2.69	0.224	—
20	2.10	0.224	0
40	1.68	0.224	0.01
60	1.26	0.224	0.03

matter in the mixture with cow manure was changed from 20 to 60%. The retention time in the first experiment was 10 days and in the others 20 days. The thermophilic digestion in 55°C was chosen first.

The results of the experiments showed that the higher the percentage of cotton stalks in the mixture, the lower the gas production. Based on calculation of digestion efficiency of the mixture of cotton stalks and cow manure (Table 5) it can be concluded that the efficiency of the digestion of the cotton stalks is nearly zero even after a long retention time.

XII. MESOPHILIC ANAEROBIC DIGESTION — SIMPLE BATCH DIGESTION

We are now operating two sets of digesters. One set of mesophilic 4 ℓ working volume digesters with cotton stalks mixed with low percentage of cow manure (5% of the cotton solids) with and without addition of nitrogen, and inoculated with digested slurry. The other experiment is a 15 ℓ vessel with 2.8 kg cotton stalks and 12 ℓ of liquids (water and inoculum) which are circulated for 2 min every 20 min through the solid fraction. The gas produced in each vessel is recorded separately. Figures 4 and 5 show the setup.

Cotton stalks of this year harvesting were first chopped dry in the Hubbert blender. Then water was added up to approximately 9% and the mixture was blended again for half an hour and kept in refrigerator overnight. To the mixture of 12 kg, 475 g of fresh manure (12% solids) from the cow barn were added and mixed for another half an hour. 3640 kg were put in 5 ℓ plastic containers and 350 g of active mesophilic inoculum of anaerobic digestion of slaughterhouse waste material was added to each vessel. Samples were taken for chemical composition (see Table 6). 0.5 g of ammonium nitrate in water were added to three of the six digesters.

All digesters were put in a thermostatic water bath with circulated water at 35°C without stirring or mixing. Outlets were connected to simple water displacement gas meter and gas production was measured daily for 150 days. Samples of biogas were taken during the experiment for gas analysis in a Varian Gas Chromotograph.

Experiments were discontinued after 150 days although some gas production was noted, especially in the digesters where nitrogen was added, but this gas production was very slow and not significant to the whole experiment.

When experiment stopped, composition of biogas was examined and the digested slurry from this long experiment was examined in our laboratory for the chemical composition. The chemical composition is presented in Table 7. Based on the organic material content charged and discharged to the digesters the efficiences of the process were calculated (Table 8) and compared to the total gas production recorded (in Figures 6 and 7).

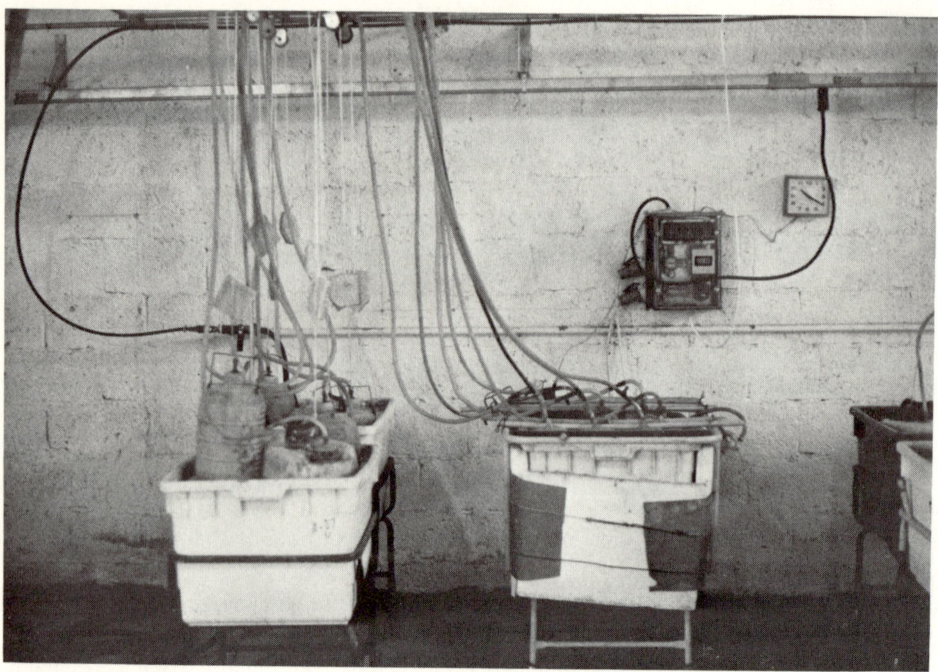

FIGURE 4.   The batch mesophilic anaerobic digestion system (of 4 ℓ working volume) of cotton stalks.

FIGURE 5.   The batch mesophilic anaerobic digestion system (of 10 ℓ volume) of cotton stalks, in which the liquids are circulated.

Table 6
**CHEMICAL COMPOSITION OF COTTON STALKS (RAW MATERIAL) AND
OF THE MIXTURES AT STARTUP OF THE ANAEROBIC DIGESTION
EXPERIMENTS**

	pH	Total solids (%)	Ash (%)	Volatile solids (%)	Total nitrogen (g/ℓ)	Ammonia (g/ℓ)	Phosphate (g/ℓ)	Total volatile acids		
								(g/ℓ)	OM/N	AOM/N
Cotton stalks	—	80.35	3.32	76.03	—	—	—	—		
Mixture without nitrogen	7.65	9.72	1.11	8.61	1.61	0.20	1.01	—	53.5	41.7
Mixture with nitrogen	7.50	9.76	0.87	8.89	1.88	0.25	1.13	1.27	47.3	36.9

Note: Mixture ratio is 95:5 (% of dry matter) cotton stalks and cow manure.

Dry cotton stalks from the last harvest were broken to size smaller than 10 cm (4 in.). 300 g of this material (241.05 g dry material) were put into a stainless steel net of 6 ℓ in volume and the net was put into the vessel of 10 ℓ in volume (see Figure 5). 9 ℓ of warm water (60°C) were added to cover the cotton stalks and the vessel was closed and heated, by 60°C water circulation, without any circulation of the liquids inside the digester. No gas production was noted. After 48 hr, 250 mℓ of inoculum from slaughterhouses digested slurry was added and still no liquid circulation operated. After another 24 hr liquid circulation (including 12 ℓ liquids from slaughterhouses digestion experiment) was operated for 2 min every 15 min throughout the experiment. Since the heating system was broken after 45 days, the experiment was not continued further.

XIII. DISCUSSION

In order to examine the potential of cotton stalks for mesophilic (37°C) methane fermentation, "long time batch" experiment (5 months) were carried out, with and without nitrogen supplementation. The digesters were initially inoculated from a mesophilic (37°C) slaughterhouse waste digesters. No significant difference was found in both cases, with or without addition of nitrogen after 150 days of fermentation (Table 8). The most striking result of the experiment was the high (about 30%) organic matter metabolized to methane value which was much higher than the expected and reported elsewhere.[31] The specific characteristics of the cotton growth on peat soil in the Hulla Valley,[18] in addition to the very long time of digestion which allows to achieve a practically total degradation of the insoluble cellulosic material present in cotton stalks, may be the direct causes of the high yield of degradation of volatile solids. Another cause which justified this phenomena may be the use of an adapted culture of microorganisms from our own experiments with slaughterhouse waste, which has relatively high percentage (approximately 15%) of wheat straw. The pH was almost constant (7.5) during the entire experiment in both cases, and was similar to the pH values achieved by mesophilic digestion of slaughterhouse wastes.

Most of the biogas was formed during the first third of the process (86% for the control and 81% for the nitrogen supplemented digesters) at a stable rate of 0.35 ± 0.01 ℓ/ℓ day for the control and 0.36 ± 0.08 ℓ/ℓ day for the nitrogen fermenters, respectively. The biogas productivity of the overall process was also similar in both cases (0.23 ℓ/gVS fed and 0.25 ℓ/gVS fed for control and nitrogen fermenters respectively). It was noteworthy that the CH_4/CO_2 ratio in the biogas was very high in both fermentations (more than 60%

Table 7
CHEMICAL COMPOSITION OF DIGESTED SLURRY AFTER 150 DAYS OF BATCH MESOPHILIC ANAEROBIC DIGESTION WITH AND WITHOUT ADDED NITROGEN

	pH	Total solids (%)	Ash (%)	Volatile solids (%)	Ammonia (g/ℓ)	Total nitrogen (g/ℓ)	Phosphate (g/ℓ)	Total volatile acids (g/ℓ)	Conductivity (mmho/cm)
Mixture without nitrogen	7.68	4.62	0.52	4.11	0.09	1.27	0.40	0.05	7.51
Mixture with nitrogen	7.69	4.05	0.56	3.51	0.18	1.20	0.46	0.11	8.85

Note: 3 replicates for each treatment. For mixture composition see Table 6.

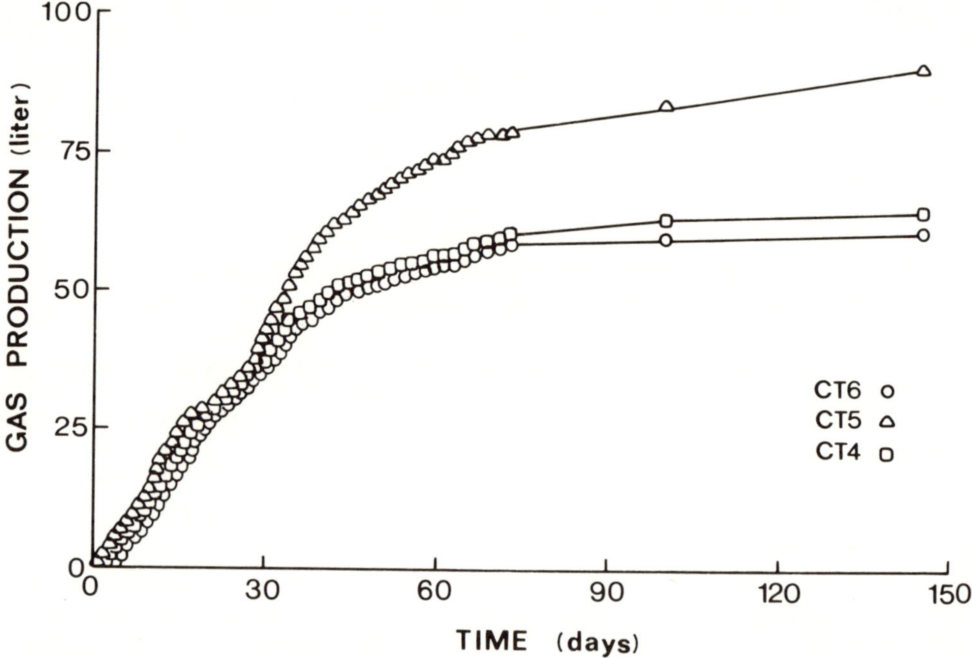

FIGURE 6.   Accumulated gas production of the batch mesophilic anaerobic digestion of cotton stalks (without added ammonia) during 150 days.

of the gas was CH_4, see Table 8). We can partially attribute this result to the relatively high (15 to 25%) intracellular soluble material found in the cotton stalks grown in the Hulla Valley. Obviously the effect of the long time of fermentation in addition to the stabilization of pH (7.5) contribute to the solubilization of the CO_2 formed. Furthermore, the probably recomsumption of the CO_2 by the methanogens, as substrate for growth due to the long period of gas evacuation after the 45 days of fermentation, may also explain the particularly high CH_4/CO_2 ratio in the biogas.

The very low concentration of volatile fatty acids during the entire process (see Table 7) was an indication of a stable fermentation, and shows a balanced interaction between the

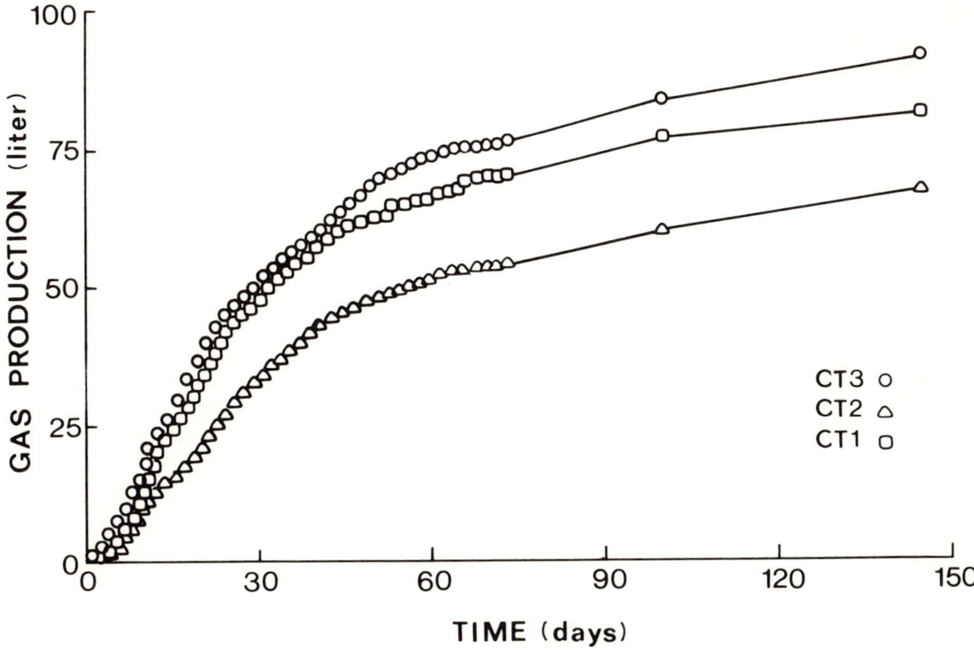

FIGURE 7.   Accumulated gas production of the batch mesophilic anaerobic digestion of cotton stalks with 0.5 g added ammonium nitrate during 150 days.

Table 8
SUMMARY OF THE RESULTS OF THE BATCH MESOPHILIC ANAEROBIC DIGESTION OF COTTON STALKS WITH AND WITHOUT ADDED AMMONIUM NITRATE (AFTER 150 DAYS OF DIGESTION IN 4ℓ WORKING VOLUME)

	Control	With nitrogen
Accumulated gas producted (ℓ)	72.3 ± 16.6	79.7 ± 11.7
Efficiency[a] (%)	27.7	30.5
Biogas production (ℓ/g VS fed)	0.23	0.25
CH_4/CO_2 in biogas[b]	1.64	1.79
First third gas production[c] rate (ℓ/ℓ day)	0.35 ± 0.01	0.36 ± 0.08

[a]   Calculated from the formula (1.2 V/VSo) × 100, where: V = accumulated gas production (ℓ); 1.2 = estimated biogas density (kg/ℓ); VSo = amount VS at startup (kg).

[b]   Percentage ratio.

[c]   Gas production rate during the first 45 days of fermentation.

three major groups of bacteria capable to degraded organic matter to methane: fermentative, H_2 producing acetogens and methanogens.

Electrical conductivity was slightly higher in digesters with added nitrogen, as expected (8.85 mmho/cm compared to 7.51 mmho/cm). These values shows a higher content of soluble salts in the slurry which were in correspondence to the fact that almost all the ash content in cotton stalks was soluble. In spite the relatively high ratio of available organic matter to nitrogen in the substrate mixture (41 in the control compared to 37 in the nitrogen treated fermenters), good results were achieved, as described above. Apparently, the cause

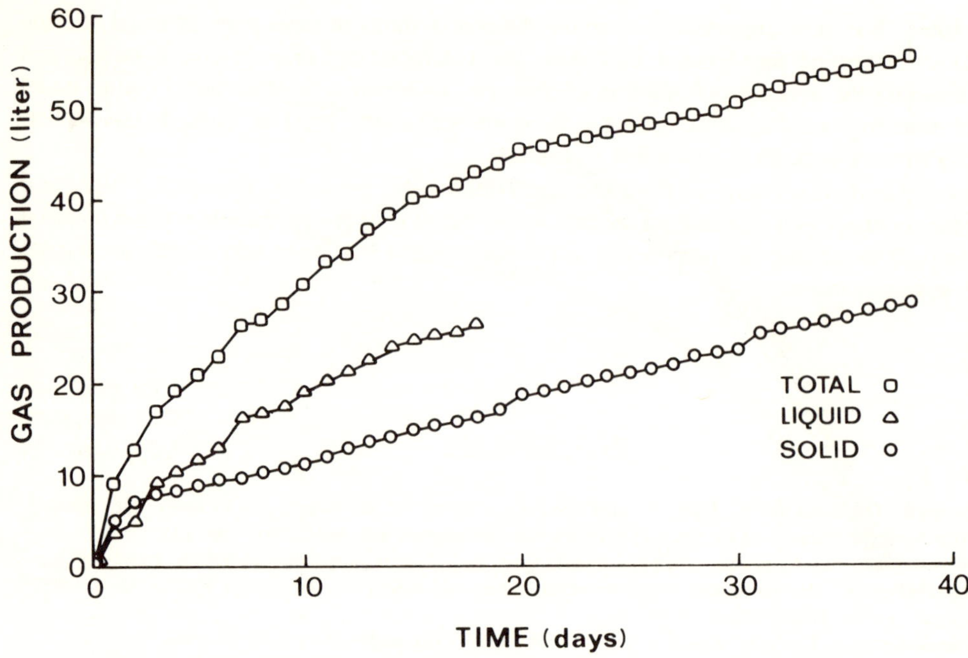

FIGURE 8.   Accumulated gas production of the batch mesophilic anaerobic digestion system with circulated liquids of cotton stalks during 45 days (until water bath control was out of order).

of that behavior was the nitrogen incorporated with the inocullum, which was in accord to the bacterial population requirements.

Results from the cotton stalks digested in the 10 ℓ volume digester with circulation of the liquids every 12 min (Figure 5) were quite similar in initial gas production rate to the digestion without circulation (Figure 8).The biogas production rate in both the liquid fraction and the solid fraction was high, but after a few days gas production from liquids stopped completely. It is difficult to come to conclusions from an uncomplete experiment but we assume no differences in initial gas production will occur when no circulation is operated. This will save in the investment for a low capital cost system. The methane to carbon dioxide ratio in the biogas was also similar.

Finally, from the results of these preliminary experiments, we concluded that the hydrolysis of the insoluble organic polymers presented in cotton stalks was the rate limiting step of the overall process of fermentation, and therefore, high retention time for continuous fermentation or long time for batch wise, would be needed to achieve the maximal potential of cotton stalks to produce methane.

XIV. CONCLUSIONS

The digestion process was performed without any mixing or liquid circulation. This is quite surprising since no lag was noted, and may be the result of using a good inoculum. The experiment with circulating the liquid every 20 min for 2 min gave similiar initial gas production. Since all cotton stalks were immersed completely in the liquid of the digestion mixture it is not different from results we reported earlier[10] with cow manure. If higher concentration of solid will be used, or cotton stalks will not be chopped so intensively, results may change.

We have not examined the effect of herbicides and pesticides used during the growth of the cotton plants, on the fermentation, but from our experiments no biological inhibition was noted. It is very important to examine the cotton stalks in other parts of Israel and the world to determine the optimal C/N ratio for anaerobic digestion of cotton stalks, and consequently the necessity of addition of nitrogen. To ensure a feasible use of cotton stalks as substrate for methane fermentation, a cheap nitrogen source must be found. In this aspect, poultry manure appears as a suitable candidate.

In Israel and other countries the amount of cotton stalks produced every year is high, and in some countries even the highest organic waste material. If this renewable organic material can be used for energy production and soil conditioning it may be of very significant impact on world agriculture.

REFERENCES

1. **Dvoskin, D., Szeskin, A., Gur, S., and Fox, S.,** A Survey of All Energy Use In Israel Agriculture, Report for the Ministry of Energy and Infrastructure, Jerusalem (in Hebrew), 1981, pp. 217.
2. Kibbutz Industries Association, NEFAH Project, Kibbutz Industries Association, Tel-Aviv, Israel, 1978.
3. **Flesenstien, G.,** Research and development on energy and agriculture, Int. Symp. Energy Agriculute, Qiryat-Anavim, Israel, March, 1983, 26.
4. **Flesenstien, G.,** Energetic aspects of growth under covers, *Hassadeh,* 64 (11), 2337, 1984.
5. **Dvoskin, D., Link, O., Zor, G., and Gazit, M.,** Survey of Green-Houses Heating Systems (A), Heshev, Tel-Aviv, Israel, 1983.
6. **Ben-Ghedalia, D. Shefet, G., Miron, J., and Dror, Y.,** Effects of ozone and sodium hydroxide treatments on some chemical characteristics of cotton straw, *J. Sci. Food Agric.,* 33, 1213, 1982.
7. **Keli, J.,** Personnal communication, 1981.
8. **Tagari, H.,** The nutritional value of digested and of fresh cattle manure, in *Utilization of Agricultural Wastes — NEFAH Project,* Research progress report No. 5, Kibbutz Industries Association, Tel-Aviv, Israel, 1978.
9. **Wise, D. L., Wentworth, R. L., and Ashare, E.,** Methane production from biomass and agricultural residues, *Ind. Eng. Prod. Res. Dev.,* 18, 151, 1979.
10. **Marchaim, U. and Criden, J.,** Research and development in the utilization of agricultural wastes in Israel for energy, feedstock fodder, and industrial products, in *Fuel Gas Production from Biomass,* Vol. 1, Wise, D. L., Ed., CRC Press Inc., Boca Raton, Fla., 1981, 95.
11. **Rymon, D., and Weinstein, M.,** A Survey of Organic Waste From Agricultural Sources of Potential Value for Recycling in Israel, Special Publication No. 114, The Volcani Center, Bet-Dagan, Israel (in Hebrew), 1978.
12. **Shafir, Z., Moaz, Y., Shaltiel, M., and Segal, I.,** A survey on the potential of plant wastes for energetic uses, submitted to the Ministry of Energy and Infrastructure, (Israel in Hebrew), 1984.
13. **Marchaim, U., Ilany, J., and Dozoretz, C.,** Preliminary study for the production of methionine-rich yeasts, Research No. 7381, Ministry of Industry and Commerce, Israel (in Hebrew), 1980.
14. **Goldstein, I. S.,** Biomass availability and utility for chemicals, in *Organic Chemicals from Biomass,* CRC Press, Inc., Boca Raton, Fla., 1981.
15. **Linden, J. C., Murphy, V. G., and Smith, D. H.,** Forage crops as chemical feedstocks, in *Bioconversion Systems,* Wise, D. L., Ed., CRC Press Inc., Boca Raton, Fla., 1984, 71.
16. **Wise, D. L., Ed.,** *CRC Series in Bioenergy Systems,* Vols. 1 to 5; CRC Press Inc., Boca Raton, Fla., 1984.
17. **Rugg, B., Amstrong, P., and Dreiblatt, A.,** Liquid fuel and chemicals from cellulosic residues by acid hydrolysis, in *Liquid Fuel Developments,* Wise, D. L., Ed., CRC Press Inc., Boca Raton, Fla., 1984, 139.
18. **Silanikov, N. and Levanon, D.,** Cotton straw composition, variability and effect of anaerobic preservation, *Biomass,* 9, 101, 1986.
19. **Marchaim, U., Perach, Z., and Kimchie, S.,** An integrated approach to the anaerobic digestion process, in *Fuel Gas Systems,* Wise, D. L., Ed., CRC Press Inc., Boca Raton, Fla., 1984, 141.
20. **Shefet, G. and Ben-Ghedalia, D.,** Effect of ozone and sodium hydroxide treatments on the digestability of cotton straw monosaccharides by rumen microorganisms, *Eur. J. App. Microbiol. Biotechnol.,* 15, 47, 1982.
21. **Ben-Ghedalia, D., Shefet, G. and Miron, J.,** Effect of ozone and ammonium hydroxide treatments on the composition and *in vitro* digestability of cotton straw, *J. Sci. Food Agric.,* 31, 1337, 1980; *J. Sci. Food Agric.,* 33, 1213, 1980.

22. **Wentworth, R. L., Ashare E., and Wise, D. L.,** Fuel gas production from animal residue. I, *Res. Rec. Conserv.,* 3, 343, 1979.
23. **Buswell, A. M.,** Anaerobic Fermentation Illinois State Water Survey, Bull No. 32, 1936.
24. **Buswell, A. M.,** in *Industrial Fermentations,* Vol. w, Underkofler, L. A. and Hickey, R. J., Ed., Chemical Publishing Co., New York, 1949, 518.
25. **McCarty, P. L.,** Anaerobic waste treatment fundamentals. I. Chemistry and Microbiology, *Public Works,* 95, 107, 1964.
26. **Hobson, P. N., Bousfield, S., and Summers, R.,** *CRC Crit. Rev. Environ. Control,* 4, 2, 1974.
27. **Stewart, D. J.,** Methane from crop-growing biomass, in *Fuel Gas Systems,* Wise, D. L., Ed., CRC Press Inc., Boca Raton, Fla., 1983.
28. **Marchaim, U.,** Anaerobic digestion of agricultural wastes. The economic lie in the effluent uses, Int. Symp. on Anaerobic Digestion, Boston Ma., 1983.
29. **Levanon, D., Dosoretz, C., Motro, B., and Kahn, I.,** Recycling agricultural waste for mushroom casing, *Mushroom J.,* 133, 13, 1984.
30. **Klinger, I. et al.,** NEFAH annual reports (In Hebrew), Kibbutz Industries Association, 1982, 1983.
31. **Kimchie, S.,** High-Rate Anaerobic Digestion of Agricultural Wastes, Ph.D. thesis, submitted to the Haifa Technion, Israel, 1984.

INDEX

A

Acclimation, 54
Acetate splitting, 58—59
Acetogenesis, 52
Acetogenic bacteria, 43, 57, 104
Acetogenic fermentation, 52
Acid formation in anaerobic digestion, 52, 146
Acid forming bacteria, 43, 52, 141
Acidogenic bacteria, 43
Acids, volatile, 57
Aerobic vs. anaerobic biotechnology, 102—103, 146
Aerobic digestion, biomass yield from, 43—44
Agricultural wastes in Israel, see also Farm wastes, 158—161
Air/fuel ratio of biogas, 11, 13
Alkali metal in livestock waste, 56
Ammonia
    anaerobic production of, 49
    buffering capacity of, 58
    concentration of, during garbage fermentation, 137—138
    inhibition of anaerobic digestion by, 58, 85
    inhibition of biogas production by, 138
Ammonium hydroxide, 158
Anaerobic bacteria, 61—62
Anaerobic biofilm, see Biofilm
Anaerobic biotechnology, 3
    aerobic biotechnology vs., 102—103, 146
Anaerobic contact floc based reactor, 62—67, 116
    biomass densities for, 63
    cell concentrations and output in, 81, 82
    kinetics of, 79—81
    solids separation in, 63—64
    wastewaters treated by, 62—63
"Anaerobic contact process", 146—147
Anaerobic digester, see also Fixed-film reactor; Methane reactor
    advantages of, 91
    assessment of performance of, 47—48
    biochemical parameters of, 48
    control actions for, 85
    disadvantages of, 92
    dynamic model of, 47
    efficiency of, 47—48
    failure of, 46—47
    full-scale, 124—125
    functioning and management of, 163
    with increased cell concentration, 125—126
    kinetic safety factor for, 76
    labor costs of, 46, 47
    loading rates for, 146
    maintenance costs of, 45—46
    process control expertise for, 46—47
    proprietary installations of, 124—125
    retained biomass, 45
    standardized parameters for, 47—48
    thermophilic, 164—165
    types of, 113—117
Anaerobic digestibility of livestock manures, 50, 52
Anaerobic digestion
    acetogenesis in, 146
    acid and methane forming stages of, 52, 146
    advances in, 92
    advantages of, 102
    batch mesophilic, see Batch mesophilic anaerobic digestion
    benefits and limitations of, 147
    biochemistry of, 146
    capital costs of, 45
    of complex wastes, kinetics of, 76—78
    of cotton stalks, 162—167
        mesophilic, 163, 167—169
        thermophilic, 163—165
    destruction of *Salmonella* and *Brucella* during, 44
    dynamic kinetic model for, 83—86
    economics of, 45—46
    energy balance of, 44—45
    fixed-film, see also Fixed-film anaerobic digestion, 101—130
    floc formation during, 61
    hydrolysis in, 146
    inhibition in, 53—60, 106—107
        from ammonia, 58
        from cations, 54—56
        from concentration, antagonism, and acclimation, 53—54
        from heavy metals, 56—57
        from hydrogen, 58—59
        from pH, 57—58
        from sulfide and sulfates, 59—60
        from toxic materials, 60
        from volatile acids, 57
    kinetics of, 104—105
    limitations of, 102—103
    of livestock wastes, 43—47
    mesophilic, of cotton stalks, 167—169
    methods of, 146—148
    microbiology of, 104—107
    nutrient requirements in, 105—106
    of paper mill effluent, 150—152
    rate-limiting step for, 50, 76
        kinetics of, 84—85
    separated phase, 52—53
    stoichiometry of, 105
    temperature for, 106
    toxicity in, 53—60, 106—107
Anaerobic digestion processes, 111—113
    comparison of, 146—148
    filter, 147—148
    parallel, 111
    phased, 112
    single, 111
    staged, 111
    start-up recommendations for, 122—123

Anaerobic digestion technology
  advances in, 48—60
    directions of innovations in, 48—49
    substrate supply and accessibility for, 49—53
  aerobic technology vs., 102—103, 146
  landmark, 162—163
Anaerobic fermentation, see Anaerobic digestion
Anaerobic fermenter, see Anaerobic digester
Anaerobic filter process, 147—148
"Anaerobic filters", 86
Anaerobic toxicity assay (ATA), 106—107
Animal feed, 157, 159
Animal husbandry, 157—158
Antagonism and synergism in aerobic system, 54
ATA, see Anaerobic toxicity assay
Attached film, see also Biofilm, 61—62
Attached-film reactor, 81—83
Automotive engines, see also Vehicular use of
    biogas
  applications of biogas to, 8—9
  dual fueling modifications for, 17—18

B

Bacteria
  acetogenic, 43, 57, 104
  acid forming, 43, 52, 141
  acidogenic, 43
  anaerobic, 61—62
  coliform, 143
  flocculation of, 61
  homoacetogenic, 104
  hydrolitic, 104
  mesophilic, 44, 106
  methane forming, 43, 52
  methanogenic, 43, 104
  pathogenic, 142—143
  psychrophilic, 44, 106
  sulfate-reducing, 59, 104
  thermophilic, 44, 106
Bacterial biofilm, see Attached film, Biofilm
Bacterial decay rate, 73
Bacterial growth, see also Microbial growth
  theories of kinetics of, 72—73
  toxic effect of, 54—55
Bacterial retention system, 48
Bacterial washout, 60—61, 87
Batch mesophilic anaerobic digestion, 169—171
Batch reactors, 48
Biochemical fundamentals of anaerobic fermenta-
    tion, 103—107, 146
Biochemical oxygen demand (BOD), 107
Biochemical parameters of digester contents, 48
Biodegradability of dairy manure, 49, 50
Bioenergy recovery and conservation in Israel,
    155—174
  anaerobic digestion, 162—163
    of cotton stalks, 163
  animal husbandry, 157
  composition of plant materials, 160—161
    cotton stalks, 161—163
    greenhouses, 156—157
    improving feed digestibility, 157—158
    mesophilic anaerobic digestion, 167—171
    thermophilic anaerobic digestion, 164—167
    waste utilization, 158—160
Biofilm, see also Attached film, 107—111, 123
Biogas, see also Diesel engine; Diesel fuel; Dual
    fueling
  air/fuel ratio of, 9, 11, 13
  automotive applications of, 8—9
  carbon dioxide content of, 19
  carbon monoxide content of, 19, 20, 24
  cetane number of, 4, 13, 14
  density of, 11, 12
  diesel engine uses of, see also Diesel engine, 1—
    40
  electric power generation with, 7—8
  engine requirements with dual fueling with, 7
  exhaust gas temperatures of, 18—19, 21
  explosive mixtures of, 14, 15
  fuel handling safety of, 14
  fuel properties and operation requirements of,
    10—15
  heat of combustion of, 10—11
  hydrogen sulfide content of, 9
  laminar flame velocity of, 14, 15
  methane content of, 10, 135—136
  octane number of, 13
  performance of, in diesel engines, 18—23
  from sanitary landfill in Egypt, see also Sanitary
    landfill, 131—144
  soot formation with, 19
  stationary application of, 4—9, 14
  sulfur content of, 13, 22
  thermophilic generation of, 133—143
  vehicular uses of, 4—9, 15
Biogas composition, influence of, on full load be-
    havior, 19
Biogas cylinders in garbage truck, 32, 34
Biogas production
  from anaerobic digestion, 163
  from cotton stalks, 163—173
  from garbage, 133—143
    daily rate of, 138, 140, 141
    inhibition of, by ammonia, 138
    relation of, to concentration of solids, 138, 139
    with sewage sludge added, 135—143
  from paper mill effluent, 151, 152
  in thermophilic anaerobic digester, 164—165
Biogas production rate in thermophilic digester,
    164—166
Biogas treatment methods for use in diesel engines,
    9—10
Biological growth yields from fermentation of
    wastes, 49—50
Biological oxygen demand (BOD), 107
  bacterial cell mass per unit mass of, 50
  in kinetic models, 77—78
  reductions in, 44
Biological solids retention time (BSRT), 44, 48

Biomass conversion, 158, 159
Biomass densities, 60—61, 63
Biomass residues, 158, 159, 163
Biomass retention, 60—62
    reactor designs for, see Retained biomass reactor
Biomass retention time, increasing, 103
Biomass support particles, porous, see Porous support particles
Biomass washout in start-up period, 123
Biomass yield from aerobic digestion, 43—44
Biomethanation, predictive model of, 48
Biotechnology, 3
    aerobic biotechnology vs., 102—103, 146
BOD, see Biological oxygen demand
Brake specific fuel consumption (BSFC), 18—20, 23
Brucella, destruction of, 44
BSFC, see Brake specific fuel consumption
BSRT, see Biological solids retention time
Buffering capacity of ammonia, 58
Buswell-Mueller formula, 105
Butyric acid, 137

C

"Cabutz", 164
Cairo, Egypt, municipal solid waste in, see also Garbage, 131—144
Carbon dioxide content
    in biogas, 19
    in municipal solids waste, 133
Carbon monoxide content of biogas, 19, 20, 24
Carbon monoxide at start of diesel fuel injection, 20, 24
Carrier Assisted Sludge Bed Reactor (CASBER), 116
CASBER, see Carrier Assisted Sludge Bed Reactor
Cation inhibition, toxic effect of, 54—56
Cell concentration
    in continuous stirred-tank and contact reactors, 81, 82
    increased, reactors with, 125—126
Cell mass balance of continuous stirred-tank reactor, 74
"Cell support systems", 72
Cellulose
    acid hydrolysis of, 161
    annual production of, 160
    digestibility of, 50—51
    effective utilization of, 161
    as energy source in ruminant nutrition, 157—158
    enzymatic hydrolysis of, 161
    molecular structure of, 160
Cellulose content of cotton stalks, 161
Cellulose decomposer changes, 142
Cellulose materials, composition of, 160
Cellulose utilization pretreatment, 126
Cellulosic fraction of mature plants, 161
Cetane number (CN) of methane and biogas, 4, 13, 14

CH_4, see Methane
Chemical oxygen demand (COD), 107
    removal efficiency, 51
        of fluidized-bed fixed-film reactor, 71
        in kinetic models, 77—78
        of upflow anaerobic sludge blanket reactor, 151, 152
"Clarigester", 64
CN, see Cetane number
CNG, see Compressed natural gas
CO, see Carbon monoxide
CO_2, see Carbon dioxide
COD, see Chemical oxygen demand
Compressed natural gas (CNG), 7
Conservation of energy
    in animal husbandry, 157
    in greenhouses, 156—157
Continuous stirred-tank reactor (CSTR), 48
    with biomass recycle, 48
    cell concentration and output in, 82
    cell mass balance of, 74
    fermentation of dairy manure in, 87—89
    hydraulic retention time for, 45, 75
    kinetics of, 74
    substrate mass balance of, 74—79
    substrate removal curve for, 76
    substrate utilization efficiency of, 75—76
    for treatment of domestic sewage sludge, 103
Cotton stalks
    anaerobic digestion of, 163—173
        mesophilic, 166—169
        thermophilic, 165—167
    as animal feed, 159
    chemical composition of, 167, 169
        differences in, from wheat straw, 161—162
Cotton straw, 158, 159
Cow manure
    alkali metal concentrations in, 56
    carbon to nitrogen ratios of, 49
    kinetics of fermentation of, 78
    mixed with cotton stalks, biogas production from, 167
    separated phase anaerobic digestion for, 52—53
CSTR, see Continuous stirred-tank reactor
Cylinder pressure development with dual fueling, 20—21

D

Dairy manure
    biodegradability of, 49
    fermentation of, 78, 87—89
Data acquisition subsystem for pilot digester plant, 30—31
Detonation, 7
Detonation noises in dual fueling, 21
Diesel engine, 1—40
    dual fueling in, 5, 15—25
        cylinder pressure development with, 20—21
        detonation noises in, 21

engine modifications for, 15—18
low pressure, 16
modifications for, 15—18
    cost of materials for, 33, 39
performance, 18—25
practical experiences with, 23—24
schemes for, 16
fuel properties and operation requirements, 10—15
governor for, 7
maximum/minimum, 8
Illapel project, 23, 25—28
injection system for, 7
with meso- or thermophilic digester, 4—5
municipal garbage collection truck, 31—39
performance analysis of
    with biogas, 18—23
    comparative, 22
    full-load, 18—19
    summary of, 22—23
pilot plant, 27—33
stationary and vehicular application, 4—9
torque curves of, 8, 9
use of biogas in, see also subtopics hereunder, 1—40
    theoretical approach to, 4
    treatment methods for, 9—10
    usable energy for, 4, 6
Diesel fuel
  carbon monoxide content of, 20, 24
  for ignition, 4, 9
  injection point for, 20, 23
    smoke and carbon monoxide at, 20, 24
  properties of, 10, 11
  saving, 5—7
  soot numbers of, 20, 24
Diesel oil, see Diesel fuel
Diesel TOTEM with meso- or thermophilic digester, 4, 6
Diffusional limitations of microbial growth, 83
Digester, see also Fermenter
  fixed-film or flocculated bacteria, 61
  mesophilic, see Mesophilic digester
  motor/generator and, installation of, 25—27
  plug-flow, see Plug-flow digester
  thermophilic, see Thermophilic digester
  tower, 117
Digester kinetics, 72—86
    development of a dynamic model, 83—86
    model for microbial growth and substrate removal, 72—79
    retained biomass reactor kinetics, 79—83
Digester plant, pilot, 25—31
Digestion, see also Fermentation
  anaerobic, see Anaerobic digestion
  of farm wastes, continuous, methane production by, see also Farm wastes, 41—100
Digestion characteristics of manure, 51—52
Disc-bladed turbines, 89—90
Dorr-Oliver anaerobic "clarigester", 64

Downflow stationary fixed-film reactors, 67, 68, 115
  for farm wastes, 86
"Dry anaerobic digestion", 126
"Dry fermentation", 52
Dual fueling, see also Biogas, Diesel Fuel
  of automotive engines, 17—18
  of diesel engine, see Diesel engine
  of garbage truck, see Garbage truck
  heating value of, 11, 12

E

Effluent discharge for pilot digester plant, 29
Egypt, biogas from sanitary landfill in, see also Sanitary landfill, 131—144
Electric power generation with biogas, 7—8
Energetic replacement ratio of gas, 7, 26
Energy
  cellulose as source of, 157—158
  conservation of, 156—157
  mixing, adequate, 44—45
  recycling of, in animal production, 157—158
Energy balance of anaerobic digestion, 44—45
Engine, stationary
  modifications for dual fueling, 15—18
  use of biogas in, 4
Engine efficiency vs. output power, 26
  cost of materials for, 33, 39
Engine requirements for dual fueling, 7
Environmental noise, curbing, 5—7
Enzymatic hydrolysis of cellulose, 161
Escherichia coli, 143
Exhaust gas contamination, 4—7
Exhaust gas of diesel fuel, 18—21, 32
Expanded bed reactor, 115—118
Explosive mixtures of biogas, 14, 15

F

Farm wastes, methane production by continuous digestion of, see also specific types of wastes and manures, 41—100
  anaerobic digestion of, 43—47, 91—92
    biological growth yields, 49—50
  design of retained biomass reactor, 86—92
  digester kinetics, see also Digester kinetics, 72—86
  digester performance, 47—48
  methane fermentation, 43
  methane reactor process design, 48—72
    biological considerations for biomass retention, 60—62
    fixed-film, 67—72
    floc, 62—72
    inhibition and toxicity, 53—60
    substrate supply and accessibility, 49—53
  retained biomass reactor for, 86—91
  utilization of, in Israel, 158—160

Fatty acids
  analysis of, 137
  long-chain, 77
  methane fermentation inhibition by, 57
  volatile, 123, 137, 152
FBFF reactor, see Fluidized-bed fixed-film reactor
Feedstock, see Farm wastes
Fermentation, see also Digestion
  acetogenic, 52
  anaerobic, see Anaerobic digestion
  biological growth rates from, 49—50
  "dry", 52
  methane, 43
  methanogenic, 52
Fermentation subsystem of pilot digester plant, 28—
    29
Fermenter, see Digester
Fick's law of molecular diffusion, 109
Fixed-bed reactor, 113—115, 117, 118
Fixed-film anaerobic digestion, 101—130
  design and utilization, 119—125
  microbial and biochemical fundamentals, 103—
    107
  modeling kinetics of anaerobic biofilm, 107—111
  processes, 111—113
  reactor types, 113—119
Fixed-film reactor, 67—72
  design of, 119—121
  expanded and fluidized bed, 69—72
  fluidized-bed, see Fluidized-bed fixed-film reactor
  with porous biomass support, 72
  start-up recommendations for, 122—123
  stationary, 67—68, 86, 115
  substrate utilization in, 117
  suspended growth reactor vs., 103
  types of, 113—117
  wastes treatable by, 123—124
Flammability ranges of gas/air mixtures, 15
Floc based reactor, 62—67
Flocculation as means of biomass retention, 61
Flocculent reactor, kinetics of, 81—83
Floc formation during anaerobic digestion, 61
Floc settling rates, 61
Fluidized-bed fixed-film (FBFF) reactor, 69—72
  advantages of, 71
  bed expansion in, 71
  characteristics of, 118
  chemical oxygen demand removal in, 71
  for farm wastes, 86
  features of, 115
  film thickness in, 71
  minimum fluidizing velocity in, 69—70
  porous biomass support particles in, 72
  solid support particles for, 69—70
  substrate utilization in, 117
  treating toxic substances with, 117
  voids ratio in, 69
Foaming problems in fixed-film reactors, 119
Fuel, diesel, see Diesel fuel
Fuel cost, 26—28
Fuel properties

of diesel oil and methane, 10, 11
  operation requirements of biogas and, 10—15
Fuel savings, 7
Fuel system
  combined, purpose of, 5—7
  dual, see Dual fueling

G

Garbage, see also Municipal solid waste
  composition of, 132
  fermentation of, 133—143
    changes in acid producing bacteria during, 141
    changes in cellulose decomposers during, 142
    changes in pH during, 137
    survival of pathogenic bacteria in, 142—143
  production of biogas from, 133—141
    with sewage sludge added, 133—134
Garbage truck
  biogas cylinders for, 32, 34
  biogas fuel for, 31—39
    motor modifications for, 32, 35
    performance of, in test-bench, 33, 37
    results of, 33—34
  dual fueling systems tested for, 32, 33, 37—39
Gas/air mixtures, flammability ranges of, 15
Gas bubbles in fixed-film reactors, 119
Gas composition of "La Feria" landfill, 10
Gas supply system modifications for dual fueling,
    16—18
Governor for diesel engine, 7
  automotive, modifications for dual fueling, 18
  maximum/minimum, 8
  stationary, modifications for dual fueling, 17
Greenhouses, conservation of energy in, 156—157

H

Heat of combustion (Hi) of biogas, 10—11
Heating value of mixture, 11, 12
Heavy metal inhibition of anaerobic fermentation,
    56—57
Heavy metal precipitation, 56
Hemicellulose, 160—161
Hi, see Heat of combustion
Homoacetogenic bacteria, 104
HRT, see Hydraulic retention time
H_2S, see Hydrogen sulfide
Hydraulic overload in fixed-film reactors, 119
Hydraulic retention time (HRT)
  for continuous stirred tank reactor, 45, 75
  for fixed-film reactor, 103
  for phases of anaerobic digestion, 52
  shortened, 60—61
Hydrogen inhibition of methane formation, 58—59
Hydrogen producing acetogenic bacteria, 104
Hydrogen sulfide, 8, 9
Hydrolitic bacteria, 104
Hydrolysis

of cellulose, acid and enzymatic, 161
of solid substrates, kinetic model for, 107—108
as step in anaerobic digestion, 146

I

Ignition, diesel fuel for, 4, 9
Ignition ranges and temperatures of methane, 13
Illapel project, 23, 25—27
Incubation temperature, increasing, 136
Inhibition
advances in understanding of, 92
toxicity in anaerobic digestion and, 53—60,
106—107
from ammonia, kinetics of, 86
Injection point of diesel fuel, 20
Injection system of diesel engine, 7
stationary, 17
Innoculum, start up quantity for, 122—123
Intake system modifications for dual fueling, 15—
17
Ionic strength, effect of, on microbial absorption,
62
Israel
agricultural waste in, 158—160
bioenergy recovery and conservation in, see also
Bioenergy recovery and conservation in Israel, 155—174

K

Kinetic coefficients, 55, 77
Kinetic safety factor, 76
Kinetics
of anaerobic biofilm, 107—111
of anaerobic contact process, 79—81
of anaerobic digestion, 104—105
of complex wastes, 76—78
dynamic model of, 83—86
of bacterial growth and substrate removal, 72—73
of continuous stirred-tank reactor, 74
of dairy manure fermentation, 78
digester, see also Digester kinetics, 72—86
of flocculent or attached-film reactor, 81—83
growth, of attached bacterial films, 62
of hydrolysis of solid substrates, 107—108
Monod model for, 73, 84
of plug-flow reactor, 78—79
of retained biomass reactor, 79—83
of soluble substrate removal in biofilm, 108—110
Kinetic safety factor, 76

L

"La Feria" landfill, gas composition of, 10
Laminar flame velocity, 14, 15
Lignin, 160—161
Liquefied petroleum gas (LPG), 6—7

Livestock wastes, see also specific wastes and
manures
alkali metal concentration in, 56
anaerobic digestibility of, 50, 52
anaerobic digestion applied to, 43—47
nutrient composition of, 49, 50
volatile solids reduction in, 51
Loading rate for anaerobic reactor, 146
Loading tank for pilot digester plant, 29
LPG, see Liquefied petroleum gas

M

Manure
cow, see Cow manure
dairy, see Dairy manure
livestock, see Livestock wastes
pig, 53
Manure digestion, 51—52, 87
Manure separation, advantages of, 51
Meat packing wastewater, treatment of, 62—63
Membrane separated reactors, 126
Mesophilic anaerobic simple batch digestion, 167—
169
potential of cotton stalks for, 167—173
Mesophilic bacteria, 44, 106
Mesophilic digester, 4—6, 51
Mesophilic digestion temperature range, 163
Methane (CH_4)
cetane number of, 4, 13
fuel properties of, 10, 11
ignition and temperature ranges of, 13
octane number of, 13
Methane content
in biogas, 10, 135—136
of municipal solid waste, 133
starter effect of sewage sludge on, 135—136
Methane fermentation, 43
Methane formation in anaerobic digestion, 52
Methane forming bacteria, 43, 52
Methane production
by continuous digestion of farm wastes, see also
Farm wastes, 41—100
economic value of, 46
from municipal solid waste, 133—141
daily rate of, 138, 140, 141
relation of, to concentration of solids, 138, 139
thermophilic, 133—141
Methane reactor, see also Anaerobic digester, 48—
60
Methane yield, definition of, 47
Methanogenesis, see also Methane production, 52,
58
Methanogenic bacteria, 43, 104
Microbial adsorption, effect of pH on, 62
Microbial attachment to support surface, 62
Microbial growth, see also Bacterial growth
diffusional limitations of, 83
substrate assimilation and, 72—74
substrate removal and

basic model for, 72—79
   kinetics of, 82—83
Microbial reaction steps, rate limiting, 52
Microbiology of anaerobic fermentation, 103—107
"Minimum fluidizing velocity", 69—70
Mixing devices
   for fixed-film reactors, 119
   for particle suspension tests, 88—90
Monod model for digester kinetics, 73, 84, 109
Motor/generator pilot system, see Digester plant,
   pilot
MSW, see Municipal solid waste
Municipal solid waste (MSW) experiment, 132—
   143
   methane content in, 133
   procedure for, 133
   results and discussion of, 133—143
   Salmonella and *Shigella* count in, 133

N

NEFAH project, 156, 165
"Net energy production", 44
Nitrogen content of livestock manures, 49, 50
Nitrogen requirements of substrate, 49
Nucleation sites, 62
Nutrient balance in substrate supply, 49—50
Nutrient diffusional limitations, 61
Nutrient requirements in anaerobic digestion, 105—
   106

O

Octane number (ON) of biogas and methane, 13
Organic materials
   biogas production from, 162—163
   improving digestibility of, 157—158
Organic wastes, see also Agricultural wastes; Farm
      wastes; Livestock wastes; Manure; Wastes
   anaerobic digestion of, 146, 147
   technology of treatment of, see also Biotechnol-
      ogy, 3
Otto engine run on biogas, 6
Out-diffusion of products formed inside biofilm,
   110—111
Output power, fuel cost as function of, 28
Overloading in fixed-film reactors, 119
Ozone, effect on digestibility of cotton straw, 158

P

Paddle mixing impellers, 89
Paper mill effluent, anaerobic treatment of, 150—
   152
Papierfabriek Roermond wastewater treatment plant,
   152—153
Parallel anaerobic process, 111
Particle suspension, mixing devices for, 88—91

Pathogenic bacteria in fermented garbage, 142—143
pH
   changes in, during garbage fermentation, 137
   effect of, on microbial absorption, 62
   inhibition of anaerobic digestion by, 57—58
Phased anaerobic process, 112—113
Phase separation of anaerobic digestion, 52—53
Phasing in fixed-film reactors, 119
Phosphorous content of livestock manure, 49, 50
Phosphorous requirement of substrate, 49
Pig manure, separated phase digestion of, 53
Plant materials, composition of, 160—161
Plug-flow digester, 45, 48
   kinetics of, 78—79
Porous support particles, 72, 93
   advantages of, 72, 87, 88
   cuboid-shaped, 72
Potassium content of livestock manure, 49, 50
Pressure peaks, motor, 7, 20—21, 24
Pressure regulator modification for dual fueling, 17
Propionic acid inhibition, 59, 137
Psychrophilic bacteria, 44, 106

R

Reactor, see also Digester
   anaerobic contact, 116
   attached-film, kinetics of, 81—83
   batch, 48
   Carrier Assisted Sludge Bed, 116
   continuous stirred-tank, see Continuous stirred-
      tank reactor
   continuous with bacterial retention system, 48
   expanded bed, see Expanded bed reactor
   fixed-bed, see Fixed-bed reactor
   fixed-film, see Fixed-film reactor
      fluidized-bed, see Fluidized-bed fixed-film
      reactor
   membrane separated, 126
   methane, see Methane reactor
   plug-flow, see Plug-flow digester
   recycled bed, see Recycled bed reactor
   recycled flocs, see Recycled flocs reactor
   retained biomass, see Retained biomass reactor
   rotating bed, see Rotating bed reactor
   separated phase, 52—53
   suspended growth, 103
   upflow, see Upflow anaerobic sludge blanket
      reactor
Reactor designs
   comparison of, 117—119
   innovations in, 60—72
Recycled bed reactor, 116—119
Recycled flocs reactor, 116—118
Retained biomass reactor, 62—72
   economics of, 45—46
   for farm wastes, 86—92
   fixed-film, 67—72
   floc based, 62—67
   kinetics of, 79—83

Rotating bed reactor, 116—118
Roughage levels in ruminant diet, 50
Ruminant nutrition, cellulose as source of, 157—
158

S

Saccharification of wheat straw, 158
Safety device modification for dual fueling, 16—18
Salmonella, 44, 133, 142
Sanitary landfill, see also Garbage; Municipal solid
waste
in Egypt, biogas from, 131—144
Sankey diagram
of diesel engine with meso- or thermophilic
digester, 4, 5
of digester pilot plant, 30, 31
Sewage sludge, see also Municipal solid waste
added to garbage for biogas production, 135—
143
starter effect of, 135—136
composition of, 133
SFFR, see Stationary fixed-film reactor
Shigella, 133, 142—143
Single anaerobic process, 111
Slurry
batch mesophilic digested, 167, 169, 170
thermophilic digested, 164—166
Smoke at start of diesel fuel injection, 20, 24
Smoke reduction with LPG, 6—7
Solids retention time (SRT), long, 107
Solids separation in anaerobic contact process, 63—
64
Solubility index, 112
Soot in exhaust gases, 32
Soot formation with biogas, 19
Soot numbers of diesel fuel, 20, 24
reduction of, 33, 39
SPFB, see Suspended particle, fixed-biomass
SRT, see Solids retention time
Staged anaerobic process, 111
Start-up of anaerobic process
biomass washout during, 123
difficulties in, 119
effect of sewage sludge on, 135—136
recommendations for, 122—123
temperature for, 123
volatile fatty acids monitoring during, 123
Stationary engine, dual fueling in, 4—9, 14
carbon dioxide content in, 8
fuel handling safety in, 14
modifications for, 15—17
treatment methods for, 9—10
Stationary fixed-film reactor (SFFR), 67—68, 86,
115
Stoichiometric air/biogas ratio, 9, 11, 13
Stoichiometry of anaerobic digestion, 105
Substrate
accessibility of, for methane reactor, 50—53
assimilation and microbial growth of, 72—74

removal and microbial growth of, 72—79
solid, kinetic model for hydrolysis of, 107—108
soluble, kinetic model for removal of, 108—110
Substrate mass balance of continuous stirred-tank reactor, 74—79
Substrate removal curve for continuous stirred-tank
reactor, 76
Substrate supply
for methane reactor, 49—50
nitrogen requirements of, 49
nutrient balance in, 49—50
phosphorous requirements of, 49
utilization of
by continuous stirred-tank reactor, 75—76
in fixed-film reactors, 119
by microorganisms, kinetics of, 82—83
Sulfate-reducing bacteria, 59, 104
Sulfide and sulfate inhibition of anaerobic digestion,
50—60
Sulfide toxicity in anaerobic digestion, 106
Sulfur content of biogas, 13, 22
Support particles, porous, see Porous support
particles
Suspended growth reactor vs. fixed-film reactor,
103
Suspended particle, fixed-biomass (SPFB) reactor,
86—90, 93
Synergism and antagonism in aerobic system, 54
Syntrophic associations, 104

T

Technical University Federico Santa Maria
(UTFSM), 2
test bench at, 19
Temperature
incubation, increasing, 136
start-up, for anaerobic digestion, 123
Temperature range
for anaerobic digestion, 106
of diesel fuel exhaust gas, 18—21
for mesophilic bacteria, 106
for mesophilic digestion, 163
of methane, 13
for psychrophilic bacteria, 44
for thermophilic bacteria, 106
for thermophilic digestion, 163
Thermal control subsystem for pilot digester plant,
30
Thermal energy input, mitigating, 44
"Thermal sludge conditioning", 65
Thermophilic bacteria, 44, 106
Thermophilic biogas generation, 134—144
from cotton stalks, 164—167
Thermophilic digester
with diesel engine, total-energy plant, 27—31
operating conditions of diesel TOTEM with, 4, 6
Sankey diagram of diesel engine with, 4, 5
Thermophilic digestion temperature range, 163

Thermophilic incubated fermenter, experiment with,
133—143
Thermophilic methane production, 133—141
Total-energy motor (TOTEM)
available thermal energy for, 26, 27
diesel, 4, 6, 13
Total-energy pilot plant, 27—31
Total-energy system, 4
TOTEM, see Total-energy motor
Tower digester, 117
Toxic agents, 53—54, 60
Toxicity
in anaerobic digestion, 53—60, 106—107
of bacterial growth, 54—55
influence of acclimation on, 54
influence of cation inhibition on, 54—56
influence of complex formation on, 54
sulfide, 106
Turbines, disc-bladed, 89—90

U

Upflow anaerobic filters, 64—67
Upflow anaerobic sludge blanket (USAB) reactor,
116—117, 145—153
characteristics of, 118, 149—150
chemical oxygen demand removal with, 151, 152
paper mill effluent treatment with, 150—152
schematic of, 149
start-up for, 123, 150
substrate utilization in, 117
Upflow stationary fixed-film reactors, 67—68
USAB, see Upflow anaerobic sludge blanket
UTFSM, see Technical University Federico Santa
Maria

V

Vehicular use of biogas, see also Automotive
engines
fuel handling safety in, 15
general considerations in, 4—5
purpose of, 5—7
safety and economy in, 8—9
typical installation for, 18
Voids ratio, 69—70
Volatile acids inhibition of anaerobic digestion, 57
Volatile fatty acids, 123, 137, 152
Volatile solids (VS) reduction for livestock wastes,
51
Volumetric load, dependence on yield coefficient,
120—122
VS, see Volatile solids

W

Wastes, see also Agricultural wastes; Farm wastes;
Livestock wastes; Manure; Organic wastes
alternative, for feeding animals, 157—158
complex, kinetics of aerobic digestion of, 76—78
treatable by fixed-film anaerobic reactors, 123—
124
Waste waters treated by anaerobic contact, 62—63,
147
Wheat straw, 158, 161—162

Y

Yield coefficient, dependence of volumetric load
on, 120—122